Contents

FOREWORD i

ACKNOWLEDGEMENTS iii

PART I – PRINCIPLES

Chapter 1	**Introduction** 1	
	What is waste management planning? 1	
	The impact of the Environment Agency 3	
Chapter 2	**Development of waste management planning** 5	
	The Control of Pollution Act 1974 5	
	Changes under the Environmental Protection Act 1990 6	
	Section 50 provisions 6	
	The Framework Directive on Waste 7	
	The Waste Management Licensing Regulations 1994 7	
	Waste management planning and waste movements to or from other countries 8	
	The Environment Act 1995 9	
	Development plans 10	
	Interaction between development planning and waste management planning 11	
	The role of waste management planning 12	
	The waste management plan 13	
	Co-operation between authorities 13	
Chapter 3	**The process of waste management planning** 15	
	Aggregation and disaggregation 15	
	Co-ordination and consultation 15	
	Co-ordinating waste management planning 15	
	Consultation 18	
	Commercially confidential data and level of detail 24	
Chapter 4	**The waste management plan** 25	
	Summary of the plan's contents 25	
	Consistent sequence 25	
Chapter 5	**The investigation – background and issues** 33	
	Legal basis for the investigation 33	
	Producers of hazardous waste 34	
	Frequency and nature of the investigation 35	
	Scope of the investigation 36	
	Costs and throughputs: confidentiality of data 37	
	Other sources of information 38	
	Non-controlled wastes 38	
	Waste arisings – weights and volumes 38	

	Area of investigation 38
	Who should carry out the investigation? 39
Chapter 6	**Survey of waste arisings – estimates and their improvement** 41
	Identification 41
	Dependent variable 43
	Classification 44
	Waste factors and estimation 45
	Improving the estimates 46
	Action to resolve difficulties 47
Chapter 7	**The survey – planning, pitfalls and presentation** 51
	Definitions 51
	Planning the survey 51
	Regional co-operation and approaches to the survey 51
	Objectives of the survey 52
	Incorporating other visits into the survey 53
	Accuracy and sample size 54
	Coverage and data collection method 56
	Likely errors 56
	Data manipulation 56
	Data coding, data entry and data checking 56
	Data analysis 57
	The importance of good form design 57
	Feedback to the respondents 57
Chapter 8	**Future developments** 59
	Planned waste management operations 59
	Future waste arisings 59
	The effects of future legislation 61
	Changes in other areas' wastes capacities 61
Chapter 9	**Costs and benefits** 63
	Resources and burdens 63
	Benefits to industry 64
	Improving waste management and waste minimisation 65

PART II – PRACTICE

Chapter 10	**Waste movements study** 69
	Licensed facilities 69
	Sites exempt from waste management licensing 70
	Licensed and exempt sites 70
	Waste flows 71
	Flow charts 71
	Using the results – balancing supply and demand 72
	Improving records of waste received at waste management facilities 72
	Timing of the waste movements study 74
Chapter 11	**Census of waste management facilities** 75
	Data sources 75
	Data according to type of site 75

Chapter 12 **Key waste arisings and their estimation** 77
Types of waste to be reported separately 77
A1-A5 Household wastes. 78
B Commercial waste 81
C Industrial waste (excluding D: construction and demolition waste). 82
D1 Construction and demolition wastes (excluding
excavated materials) 83
D2 Excavated materials (engineering spoil) 84
D3 Contaminated soils 85
G Incineration residues 85
H Health Care wastes 86
I Sewage sludges 86
R Mine and quarry wastes 87
S Agricultural wastes 89

Chapter 13 **Selecting industrial and commercial firms for the survey** 91
Profiling the area. 91
Sub-dividing the population 91
Reference numbers for waste producers 92
Separating industrial, commercial and other waste
producers within SIC(92) 92
Creating the separate groups. 93
Sample sizes 93
Data collection methods 94
Deciding the approach. 95
Selecting firms to survey by personal visit. 98
Selecting firms to be surveyed by postal questionnaire. 99
Creating the final listings of firms to be surveyed 100

Chapter 14 **Guidelines for survey by personal visit** 101
Approaching firms to be visited 101
Designing the data forms 102
The visit 103

Chapter 15 **Guidelines for postal surveys** 107
General administration 107
Questionnaire content and design 108
Mailing 108
Dealing with returned questionnaires 109
Increasing the response 110
Storing returned questionnaires: archiving data 110

Chapter 16 **Data management, checking and validation** 113
Computer is better for managing large datasets 113
Spreadsheet or database? 113
Database design 113
Data coding 114
Data checking and validation. 114
Deriving weight of waste from volume of waste 115
Using local knowledge to verify results 117

Appendix A	**The Controlled Waste Regulations 1992 (SI 1992 No 588): Waste to be treated as industrial waste**	121
Appendix B	**SIC codes for industrial and commercial waste producers**	125
Appendix C	**Suggested format for personal visits questionnaire**	131
Appendix D	**Example postal questionnaire for small firms**	135
Appendix E	**Example postal questionnaire for large firms**	147
Appendix F	**Data coding frame and form**	161

List of Figures

Box 2.1	The WRA's planning obligations under EPA90.	6
Box 2.2	The waste framework directive's provisions for waste management planning	7
Box 2.3	A summary of PPG23 guidance on the content of the waste local plan	11
Box 3.1	The waste regulation authority's EPA90 duty to consult statutory and corporate bodies.	18
Box 3.2	EPA90 publicity and public consultation for the draft waste management plan	19
Box 4.1	Information to be included in the EPA90 plan	26
Box 5.1	Waste planning objectives: paragraph 4(3) of Schedule 4 to the Waste Management Licensing Regulations 1994	34
Box 12.1	Wastes defined and prescribed as household waste	79
Box 12.2	Wastes defined and prescribed as commercial waste or not commercial waste	81
Box 12.3	Definition of industrial waste.	82
Box 12.4	Minerals types used by minerals planning authorities to summarise types of development	88

List of Tables

Table 6.1	Comparison of the census of employment with the business database for use as a sampling frame	42
Table 6.2	Illustrative section of SIC 1992.	45
Table 7.1	Data gathering objectives and rationale.	53
Table 11.1	Data items to be collected in the census.	76
Table 12.1	Typical land area required for spreading wastes from different livestock	90
Table 13.1	Area industry profile of information from the census of employment	91
Table 13.2	Comparison of the two main survey methods	94
Table 14.1	Data to be collected about each type of waste, and the reason for its collection	103
Table B1	SIC(92) Divisions to be classed and industrial waste producers	125
Table B2	SIC(92) Divisions to be classed as commercial waste producers	126
Table B3	SIC(92) Divisions to be treated separately from the survey	127
Table F1	Coding frame	161

Foreword

Waste management planning has for too long been the 'cinderella activity' of waste regulation.

Waste regulation staff are responsible for protecting the wider environment from the effects of waste management operations, licensing waste management facilities and ensuring they operate within the conditions of those licences, as well as many other regulatory duties in relation to waste. Tasked, as they are, with this vital work, too little thought has been given to and too few resources have been directed at waste management planning.

When waste management planning has been done, the production of a waste management plan has often been seen as the sole objective and the only use to which the information will ever be put. Waste management plans have largely been founded on inadequate information arising from one-off surveys of waste producers that cannot be usefully related either to surveys in neighbouring areas or to data collected to inform previous plans.

However, there is a combination of existing and new demands now being made (or soon to be made) of waste management planning:

- the requirements of the Framework Directive on Waste,
- the forthcoming waste strategy,
- the need to inform development plans and regional planning guidance,
- the new Environment Agency.

Thus, waste management planning is at a crossroads of development. Both the legislative requirements and the organisation that is required to meet them are embarking on a period of fundamental change. At the same time, the level of detail and accuracy required of information in order to make sensible decisions has increased by an order of magnitude and the need now is not just to collect local information, influencing local decisions, but also the provision of national information, impacting on national policies.

Waste management planning is no longer an option. It is an essential part of waste regulation and provides the framework for decision-making in waste management that optimises both environmental protection and costs.

This guidance sets out how these issues should be addressed. It marks a step change in the concept of waste management planning and should serve equally well both as guidance to waste regulation authorities and to provide an initial framework of best practice within which the Environment Agency can operate.

Terry Coleman
Wastes Technical Division

Acknowledgements

The Department of the Environment is grateful for the assistance of the following in the production of this guidance:

Dr R Pocock, Ms B Leach and Mr P J Edwards of MEL Research, who provided much of the detailed methodology for the industrial and commercial waste surveys and comments on the text.

Dr Ian White, Devon County Council, for his work on the original consultation draft

Those waste regulation officers and waste management planners who provided valuable additional comment on the draft text.

PART I – PRINCIPLES

Chapter 1 Introduction

What is waste management planning?

1.1 Waste management planning has all too often been misunderstood or limited by preconceptions as to its purpose. Here it means planning in its broader sense – assessing the consequences of different options, determining the best and setting out the actions required to secure this objective. It includes, but encompasses much more than, the preparation of a waste management plan. Thus, waste management planning provides the underlying foundation for sound decision-making in waste management. It entails the collection, analysis and periodic presentation of information relevant to the management of waste in an area. This will include information on:

> the quantities of the various types of waste

> their sources

> the costs of different methods of managing these wastes, and

> their environmental effects (including the impacts of transport).

1.2 Waste management planning is an iterative process, requiring frequent reassessment of the position as new data constantly update the information already held on the quantities, costs and impacts.

This guidance

1.3 This paper

>[1] explains the purpose of waste management planning under section 50 of the Environmental Protection Act 1990[2] and the Environment Act 1995

> underlines the significance and the role of waste management planning in development control planning for waste management facilities

> gives guidance on how waste regulation staff[3] might most effectively undertake the twin requirements of section 50[4]

> 1 to carry out an investigation of waste arisings and waste management facilities and

---

[1] This paper uses one of two symbols to mark a sub-paragraph.
> The > marks the sub-paragraph as one of a sequence (as here), or – occasionally – as one that extracts subordinate material to simplify the main paragraph.
> The • is emphatic: the material is set apart because it demands particular action, or at least particularly close attention.

[2] as amended by the Waste Management Licensing Regulations 1994 (SI 1994 No.1056) to incorporate the requirements of the Framework Directive on Waste (75/442/EEC as amended by 91/156/EEC)

[3] In this paper the terms waste regulation authority, WRA and waste regulation staff are sometimes interchangeable. The convention is that 'WRA' may mean **either** the authority's members **or** the authority's employees: the context usually shows which. Where the meaning is 'the authority's employees', there is the further implication 'those employees who would be expected to undertake the task under discussion'.

[4] as amended

2   to prepare a waste management plan[5], including

     a   the scope of the investigation, and how to carry it out, and

     b   how to prepare a waste management plan.

> does this in a way that allows WRAs and subsequently the Environment Agency[6] to provide the breadth and depth of information now required.

1.4 Waste management planning provides a framework for the private and public sectors to

> **take strategic decisions for the minimisation, recovery or disposal of waste in a way that safeguards the environment.**

1.5   A waste management plan essentially calculates

     1   how much waste has to be dealt with in the area

     2   how much waste the various waste management facilities can deal with adequately

     3   the differences between **1** and **2**, whether shortfall or excess.

and then goes on to show

     4   how the differences may be resolved.

1.6   In most areas of England, the collection and the disposal of household waste are the responsibilities of different authorities. In Wales, waste regulation, waste collection and waste disposal are the responsibility of the same authority. Additionally, in both England and Wales the WRA must prepare the plans, but local authorities cannot provide waste management facilities. However, all WRAs should use the waste management planning process

> to bring together waste collection authorities to discuss their recycling activities and plans, and to provide a co-ordinated direction for recycling

> to provide information to development planners about future needs for waste management facilities of different types

> to provide information to waste disposal authorities on the environmental impacts of the options for dealing with household waste

---

[5] Section 50 of the Environmental Protection Act 1990 uses (in the marginal note) the words 'Waste disposal plans'. But even as originally enacted, the section dealt explicitly with treatment as well as disposal – see s50(3)(d): so 'waste disposal plan' was already a somewhat narrow term. The Waste Management Licensing Regulations 1994 (schedule 4, paragraph 9(8)) redefined 'disposal' in section 50(3) to include recovery. They did this to conform with the Waste Framework Directive, in which the plans are called 'waste management plans'. This paper likewise calls them 'waste management plans'.

[6] The Environment Act 1995 provides for the creation of the Environment Agency for England and Wales (and the Scottish Environmental Protection Agency). On the appointed day the Environment Agency will become responsible for the functions of waste regulation authorities in England and Wales, including the waste management planning function. So far as the context allows, this paper's references to the duties of WRAs are to be read equally as references to the duties of the Environment Agency after the appointed day.

> to assist industry in its investment decisions; and

> to identify opportunities for environmentally sound waste minimisation, re-use, recover and recycling.

1.7 Thus the waste management plan is the framework for decisions on investment as well as environmental protection. It enables waste managers to deal with controlled waste in an environmentally sound and efficient way, and assists the UK to fulfil its obligation[7] to draw up waste management plans.

The impact of the Environment Agency

1.8 Under the Environment Act 1995, section 50 of EPA90 will be repealed and its detailed requirements replaced with more general provisions requiring the Environment Agency, among other things:

> to compile information on environmental pollution; and

> to carry out surveys or investigations as directed by the Secretary of State.

The purpose of this latter requirement is to ensure the Environment Agency provides the information required to inform policy initiatives and, in particular, the development and review of a statutory waste strategy, which the Secretary of State will be required to draw up.

1.9 This guidance sets out an approach to waste management planning that will supply the waste management information needs of WRAs and ensure that this information can be used effectively by the Environment Agency as a basis for its waste management planning to meet these requirements.

1.10 This requires a properly structured approach to waste management planning and particularly to surveys and investigations. Detailed guidance is given in this paper on carrying out the investigation of waste arisings, using a sampling strategy devised at the national and regional level and the new national waste database, in order to improve the accuracy and consistency of information. The survey information is cross-checked with the corresponding weight data from waste management facilities.

1.11 Waste management planning should not be driven just by the need to publish a plan or the provision of information for a waste strategy – it is a constant, iterative process that will govern the policies and provide a framework for the everyday decisions of the Environment Agency in relation to waste management. However, the use of a structured approach alone is not sufficient to meet these aims. Unless the Environment Agency uses common definitions and methods, with the assistance of suitable systems and tools, waste management planning, at best, will remain fragmented and incapable of providing an accurate picture of waste management on a wider scale.

- This guidance introduces a common means of describing wastes, the national waste classification system, which the Department intends to introduce in 1996.

---

[7] under article 7 of the Waste Framework Directive

The survey and waste minimisation

1.12　Waste regulation staff have a key role in improving waste management and waste minimisation. For this reason, the industrial survey may be used as an opportunity to provide baselines for waste minimisation and provide wider advice to industry on improving its waste management practices. Such improvements can only result from the proper integration of waste management, where practical experience helps to improve the guidance on best practice, and waste management planning is seen as a relevant framework within which to make consistent decisions that offer the best practicable option for the environment in each case.

Chapter 2 Development of waste management planning

2.1   This chapter describes the development of the legislative framework for waste management planning in England and Wales. It discusses

> the established role of the WRA in waste management planning

> the present statutory requirements for waste management planning

> waste movements to or from other countries

> new requirements affecting waste management planning to be introduced under the Environment Act 1995

> development plans and the interaction between development planning and waste management planning

> the future role of waste management planning.

The Control of Pollution Act 1974

2.2   In 1976 waste disposal authorities[8] acquired statutory obligations[9] to

- investigate
    > the types and quantities of waste arising
    > the capacity available to deal with the waste
    > the import and export of waste; and to
- draw up plans for dealing with the waste, now and in the future.

Changes under the Environmental Protection Act 1990

2.3   In 1991, the CoPA provision was replaced by s50 of the Environmental Protection Act 1990 (from here on, EPA90).

2.4   In the waste management context, the important differences between CoPA and EPA90 are that

a   CoPA enabled waste disposal authorities to provide waste management facilities[10]. Under EPA90's separation of functions, a WRA in England or Wales cannot provide waste management facilities

b   under EPA90[11], the WRA's investigation (see chapter 3 below) should enable the WRA to prevent, or minimise, pollution of the environment and harm to human health

---

[8] For waste management planning purposes the waste disposal authority (as defined under the Control of Pollution Act 1974) was the direct predecessor of the WRA.

[9] Control of Pollution Act 1974 (from here on, CoPA), s2

[10] Because statute uses 'facilities' as a catch-all term, this paper does so too.

[11] s50(1)(a)

    c   under EPA90[12] the WRA, when planning, must 'have regard to the desirability, where reasonably practicable, of giving priority to recycling waste'.

Section 50 provisions

2.5 EPA90 s50 obliges the WRA to investigate wastes and the arrangements for its management in its area and plan what arrangements might be needed in the future: see box 2.1.

The Framework Directive on Waste

2.6 Waste management planning is also subject to European Communities (EC) law. Article 7 of the Waste Framework Directive imposes obligations on the UK government to ensure that the 'competent authorities' – in this context the phrase means the WRAs and local planning authorities[13] – make waste-management plans: see box 2.2.

2.7 The UK's **waste planning** objectives under the Framework Directive are, in summary,

    a   encouraging waste reduction, waste re-use, waste recycling, and waste-to-energy

    b   ensuring waste is recovered or disposed of without harm to

        >   human health

        >   the environment

        >   local amenity

    c   achieving self-sufficiency in waste disposal: that is, being able to dispose of all controlled wastes that arise in the UK at disposal sites within the UK.

Box 2.1   *The WRA's planning obligations under EPA90*

Under s50(1) of EPA90 the WRA must
- (a) carry out an investigation with a view to deciding what arrangements are needed for the purpose of treating or disposing of controlled waste[14] which is situated in its area and controlled waste which is likely to be so situated so as to prevent or minimise pollution of the environment or harm to human health;
- (b) decide what arrangements are in the opinion of the authority needed for that purpose and how it should discharge its functions in relation to licences;
- (c) prepare a statement ('the plan') of the arrangements made and proposed to be made by waste disposal contractors, or, in Scotland, waste disposal authorities and waste disposal contractors, for the treatment or disposal of such waste;
- (d) carry out from time to time further investigations with a view to deciding what changes in the plan are needed; and
- (e) make modifications of the plan which the authority thinks appropriate in consequence of any such further investigation.

Under s50(2) of EPA90 the WRA, in considering any arrangements or modifications for the purpose of subsection (1)(c) or (e), must have regard both to the likely cost of the arrangements or modification and to their likely beneficial effects on the environment.

---

[12] s50(4)

[13] When the statutory waste strategy is produced, this will also include the government.

[14] The Waste Management Licensing Regulations 1994, in paragraph 9(2) of schedule 4, provide that any reference to waste in part II of EPA90 shall include a reference to Directive waste. Directive waste is defined in regulation 1(3) of the Waste Management Licensing Regulations 1994. Controlled waste should be taken to mean Directive waste.

> **Box 2.2** *The waste framework directive's provisions for waste management planning*

Article 3 (part):

Member States shall take the appropriate measures to encourage:
- (a) firstly, the prevention or reduction of waste production and its harmfulness.....
- (b) secondly:
  - (i) the recovery of waste by means of recycling, re-use or reclamation or any other process with a view to extracting secondary raw materials, or
  - (ii) the use of waste as a source of energy.

Article 4:

Member States shall take the necessary measures to ensure that waste is recovered or disposed of without endangering human health and without using processes or methods which could harm the environment, and in particular:
- without risk to water, air, soil, plants and animals,
- without causing a nuisance through noise or odours,
- without adversely affecting the countryside or places of special interest.

Member States shall also take the necessary measures to prohibit the abandonment, dumping or uncontrolled disposal of waste.

Article 5:

Member States shall take appropriate measures, in co-operation with other Member States where this is necessary or advisable, to establish an integrated and adequate network of disposal installations, taking account of the best available technology not involving excessive costs. The network must enable the Community as a whole to become self-sufficient in waste disposal and the Member States to move towards that aim individually, taking into account geographical circumstances or the need for specialized installations for certain types of waste.

The network must also enable waste to be disposed of in one of the nearest appropriate installations by means of the most appropriate technologies in order to ensure a high level of protection for the environment and public health.

Article 7 (part)

In order to obtain the objectives referred to in Articles 3, 4 and 5 the competent authority or authorities ...shall be required to draw up as soon as possible one or more waste management plans. Such plans shall relate in particular to:
- the type, quantity and origin of waste to be recovered or disposed of;
- general technical requirements;
- any special arrangements for particular wastes;
- suitable disposal sites or installations.

Article 7 also provides that plans may include details of:
- the natural or legal persons empowered to carry out the management of waste;
- the estimated costs of the recovery and disposal operations;
- appropriate measures to encourage rationalisation of the collection, sorting and treatment of waste.

2.8 These requirements of the Framework Directive are met by the waste management planning regime, the development control planning regime and the pollution control regimes under Part I of EPA90.

The Waste Management Licensing Regulations 1994

2.9 The Waste Management Licensing Regulations 1994[15] (from here on, the 1994 regulations)

> transpose the provisions of the Framework Directive into British law

---

[15] *The Waste Management Licensing Regulations 1994.* SI 1994 No.1056. London: HMSO: 1994. ISBN 0 11 044056 0.

> amend the statutory[16] meaning of controlled waste so that it aligns with the Framework Directive's definition of waste: the effect of this is that 'controlled wastes' now include other wastes subject to the Framework Directive

> modify[17] the plan-making functions of waste regulation authorities to take account of the Framework Directive.

Waste management planning and waste movements to or from other countries

2.10 Waste management planning will be affected by demands made on UK facilities as a result of wastes exported from other countries. This effect may be significant for some recycling operations; but less so for most other kinds of facilities, because of restrictions placed on other waste movements by international agreements, European legislation and UK government policy.

European law

2.11 Article 5 of the Waste Framework Directive (see box 2.2) requires the European Community[18] as a whole to be self-sufficient in waste disposal and emphasises the desirability for member states individually to aim at such self-sufficiency. However

> geographical circumstance may prevent complete self-sufficiency, and

> it may not be feasible for specialised installations (e.g. for the disposal of PCB) to be provided in every member state.

Waste shipments

2.12 Since 6 May 1994[19] all UK imports and exports of waste are governed by the EC Waste Shipments Regulation and the accompanying UK regulations[20]. Where waste is being moved for disposal, competent authorities in each EC member state are able to

> prohibit generally or

> prohibit in part or

> object systematically to

proposed imports of waste.

Implementation and government policy

2.13 It is UK government policy[21]

> that the UK should be self-sufficient in waste disposal

> that *exports* of waste **for disposal** be banned

> that *exports* of waste **for recovery**

---

[16] under CoPA and EPA90

[17] by paragraphs 3 and 4 of schedule 4

[18] This paper uses, according to context, 'European Communities', 'European Community', and 'European Union'.

[19] Article 4(3)(a)(i) and (b) of Council Regulation 259/93 on the supervision and control of shipments of waste within, into and out of the European Community

[20] The Transfrontier Shipment of Waste Regulations 1994 (SI 1994 No. 1137)

[21] Government policy on the transfrontier shipment of waste will be published in the UK's management plan for the import and export of wastes.

> to OECD countries should be allowed

> to non-OECD countries should be prohibited in line with Decision II/12 of the parties to the Basel Convention

> that *imports* of wastes **for disposal** should generally be banned, except where:

> the country intending to export the waste does not itself have the technical capacity and necessary facilities to dispose of the waste question in an environmentally sound manner

and that country cannot be reasonably acquire the technical capacity and necessary facilities

and the waste will be managed in an environmentally sound manner at the destination facility in the UK

> that *imports* for genuine reclamation or recovery will continue to be allowed.

2.14 Waste management planning is a key element in implementing the UK's policy for imports and exports of waste. It is also an important step towards UK self-sufficiency in waste disposal.

The Environment Act 1995

2.15 The Environment Act 1995

> transfers[22] to the Environment Agency the functions of the waste regulation authorities in England and Wales

> repeals[23] s50 of EPA90

> inserts[24] into EPA90 a new section, 44A, on the National Waste Strategy: England and Wales,

> provides in section 44A that the Secretary of State may direct the Agency to survey or investigate

> the kinds and quantities of wastes

> the recovery and disposal facilities likely to be needed

> anything else the Secretary of State wishes to know for the waste strategy

> requires[25] the Agency to compile information on environmental pollution to

> carry out its pollution control functions, or

> enable it to form an opinion on the general state of pollution of the environment.

---

[22] section 2(1)(b)

[23] by s120 and schedule 24. The repeal will take effect on a date to be specified by order: see s125(3) of the Act.

[24] by s92(1)

[25] section 5(2)

Development plans

2.16 Land-use planning for waste facilities is part of the development planning system.

2.17 Statutory development plans[26] set out the main considerations for the local planning authority[27] to use in deciding planning applications. They can also guide other responsibilities both of local government and other agencies.

2.18 In England and Wales the statutory development plans are

> **structure plans** and

> the four kinds of **local plans**[28]

    1   waste local plans[29]

    2   minerals local plans[29]

    3   district local plans

    4   national park local plans.

Structure plans and waste local plans have most relevance to waste management planning. Their significance in this context is considered in paragraphs 2.19 and 2.26 below.

The structure plan

2.19 The structure plan (or, in a metropolitan area, part I of the UDP) states the strategic policies for the development control planning of wastes facilities. These strategic waste policies must

- have regard to

  > national guidance – in Planning Policy Guidance Notes (PPGs) and Minerals Planning Guidance Notes, and

  > regional guidance issued by the Secretary of State

- ensure that waste local plans (see paragraphs 2.20 to 2.22 below) provide for waste facilities sufficient to meet the estimated need over the plan period.

The waste local plan

2.20 In England, in addition to the structure plan, the county planning authority also has to draw up a *waste local plan*: the waste local plan may be free-standing, or may be part of a joint minerals and waste local plan.[30] These plans should conform to the strategic framework in the structure plan.

2.21 For England[31], PPG23 *Planning and Pollution Control*[32] provides guidance to local planning authorities on the role and content of

---

[26] Under Part II of the Town & Country Planning Act 1990

[27] References to local planning authority in this paper should be read as including the minerals planning authority

[28] except in metropolitan areas, which have unitary development plans (UDPs). In Wales, from 1 April 1996 the new unitary authorities and national park authorities will also produce UDPs instead of structure and local plans.

[29] The minerals local plan and the waste local plan may be combined.

[30] In metropolitan areas, equivalent policies should be included in Part II of the UDP.

[31] Welsh Office Circular 26/94 in Wales

[32] DoE, (1994) PPG23 – Planning policy guidance note: planning and pollution control. HMSO. London. ISBN 0 11 752947-8

development plans, and particularly waste local plans. Box 2.3, below, summarises paragraph 2.23 of PPG23.

2.22 The waste local plan, and other development plans that include waste policies, must[33] take account of waste management plans. More particularly, the local planning authority will derive from the waste management plan its view of

> the types and quantities of waste to be dealt with
> the waste-management facilities needed to do this.

Box 2.3   *A summary of PPG23 guidance on the content of the waste local plan*

In England, the waste local plan must
1. conform with the structure plan
2. have regard to the waste disposal plans for the area
3. have regard to national and regional planning guidance
4. have regard to the Waste Framework Directive.

The waste local plan should include detailed land-use policies and proposals for wastes. These policies and proposals should
> take into account, among other things,
5. regional self-sufficiency in waste management facilities
6. their land-use impacts
7. their transport requirements and resultant impacts
8. waste minimisation and recycling policies
9. opportunities for energy recovery
10. treatment methods prior to disposal
> identify
11. spare capacity, scope for new capacity
12. broad areas of search for new capacity
> include
13. criteria for deciding a planning application for a waste management development
14. criteria for improvement in the environmental acceptability of recycling and other waste management operations.

Interaction between development planning and waste management planning

2.23 Much of the information in waste local plans will be derived from the waste management planning process. Hence, as a matter of sensible administration, the local planning authority should discuss its proposals with the WRA at an early stage in the preparation of the waste local plan.

- The WRA should alert the planning authority to this need: see paragraphs 3.17 to 3.19 below.

2.24 The waste management plan should be compatible and consistent with the development plans; they should be mutually reinforcing. Thus

---

[33] Town and Country Planning (Development Plan) Regulations 1991

> information collected by the WRA for waste management planning contributes to development planning (for its particular contribution to waste local plans, see paragraph 2.25), and

> information from development planning provides the planning framework in the waste management plan.

2.25 Waste local plans must take a variety of factors into account: box 2.3 summarises them. Waste management planning should be able to provide

> **all** the information needed for items **5, 8, 9** and **10** in box 2.3; and as a consequence

> a large part of the information and advice for **6, 7, 11** and **12**.

2.26 Waste management planning and development control planning may seem rather alike: but the similarity is deceptive – there are differences both in purpose and content. The WRA should **not** abdicate its responsibilities for

> carrying out the investigation

> deciding the best way of dealing with waste in its area.

The role of waste management planning

2.27 Waste management planning is concerned primarily with strategic aspects of reducing, treating and disposing of controlled waste. A site may seem suitable for waste management, and be so discussed in the waste management plan; but the waste management plan cannot decide the suitability of the site against **land-use** criteria. That is for the development plan to do.

2.28 Waste management planning provides the framework for dealing with controlled waste in accordance with the objectives[34] of the Framework Directive.

2.29 In deciding how to meet those objectives, the WRA will take account of

> the amounts[35] of wastes of particular types

> the facilities available, or likely to become available, in particular locations.

2.30 Waste management planning should shape future decisions on waste management facilities, and provide the framework within which those decisions may be sensibly taken.

- The WRA should derive from its waste management planning work its policies for discharging its licensing functions.

---

[34] repeated at paragraph 4 of schedule 4 to the 1994 regulations

[35] In this context, 'amounts' means 'the arisings wastes produced in, and brought into an area which require to be dealt with'.

2.31 Thus, in planning for waste management the WRA should

> consider the broadest possible range of waste management options

> decide which of them is best for the area

> settle its policies for dealing with waste accordingly.

The waste management plan

2.32 The preparation and publication of a document – the waste management plan is one of the tangible outputs from the process of waste management planning. The waste management plan must[36] include certain information: Box 4.1 on page 26 sets out the information required to be included in the waste management plan. The detailed content and recommended structure of the plan are discussed in Chapter 4[37].

Co-operation between authorities

2.33 The WRA will need to work closely with the local planning authority: for the mechanisms of co-operation with local planning authorities, see paragraph 3.17. The common objectives in this co-operation are

> to consider potential waste management facilities identified by the planning authority

> to consider the extent to which land use, amenity or other environmental factors (e.g. geological or hydrogeological limitations, or the location of housing constrain different waste management options).

2.34 Waste management planning will necessarily cover recycling: it thus provides the background for the district councils to draw up or revise their recycling plans.

- Only dialogue between the authorities will ensure that district recycling plans adequately reflect overall waste management policies: see paragraphs to 3.5 and 3.19 below.

2.35 The WRA should also liaise regularly with the waste disposal authority to exchange useful information. In particular

- the WDA will have information on waste collected by waste collection authorities, waste from civic amenity sites and some recycling activities.

- the WRA will be able to provide information on

  > future waste inputs and capacities

  > the environmental impacts of all the options for dealing with waste in its area.

---

[36] section 50(3) of EPA90

[37] Annex 1 to DoE Circular 11/94 on the Waste Management Licensing Regulations provides further guidance on waste management plans: see its paragraphs 1.35 to 1.40.

2.36 This assessment of the impacts of waste management options should provide the information the WDA needs in order to fulfil the requirement[38] to decide, in any contract dealing with waste, what, if any, conditions should be included to

a minimise pollution

b maximise recycling.

---

[38] Paragraph 19, Schedule 2 to EPA90

Chapter 3 The process of waste management planning

3.1 This chapter describes the administrative mechanisms that the WRA should adopt to carry out waste management planning. It covers

> consultation

> publication and scope of the plan

> the period to be covered by the plan.

3.2 It also deals with one of the WRA's key roles: **to co-ordinate waste management planning** and sets out how waste management planning at all levels (district, county, regional and national) may be co-ordinated effectively so that plans are

> coherent and consistent at each level and

> compatible between the levels.

Aggregation and disaggregation

3.3 Waste management planning should be organised at a regional level and generally carried out at county level; but the core dataset should be built up from district-level data. The area plan's data inputs and information outputs should be capable of aggregation and disaggregation. Aggregated data then feed regional planning and national policies; disaggregated data provide the context for district council recycling plans.

Co-ordination and consultation

3.4 This chapter distinguishes co-ordination from consultation. Co-ordination is treated here (paragraphs 3.5 to 3.19) as a managerial process, driven by practical circumstance and related to waste management planning generally; consultation (paragraphs 3.20 to 3.62) is treated as more dependent on the statutory roles and legal rights of the participants and is relevant to the waste management plan.

Co-ordinating waste management planning

3.5 For effective waste management planning the WRA should maintain a dialogue with other authorities and organisations.

3.6 In most areas, the WRA should regularly – although not necessarily frequently – discuss or liaise with

> any WRA to which waste is sent

> any WRA from which waste is received

> any NRA region whose interests could be affected by the plan: this includes regions into which ground and surface waters flow or might flow

> other pollution control authorities – that is, HMIP and local authority environmental health departments – who also have records of the plants and processes in their area that deal with waste or produce waste

> local planning authorities – who operate development control but also need information from the wastes survey

> district councils whose recycling interests are affected. This may include both those within the WRA's own area and, at the drafting stage, those in other WRAs' areas.

- Consultation with the out-area districts should be through their own WRA

> relevant industry groups, e.g. the Environmental Services Association (formerly the National Association of Waste Disposal Contractors) the British Metals Federation, the Sand and Gravel Association, the British Aggregates and Construction Materials Industry

> industries or organisations producing large volumes of waste: e.g. sewerage undertakers and electricity generators

> producers of any waste stream that may need specific consideration e.g. the health authorities

> the waste disposal authority (WDA) for the area.

The most important areas for co-ordination in waste management planning are discussed below.

Regional co-ordination

3.7 For co-ordination at the regional level, the WRA should seek to use, and strengthen, the work of officer groups within existing voluntary and statutory regional groupings of WRAs. These groupings are well placed to take account of regional development planning guidance that affects their waste management plans.

3.8 Regional co-ordination also allows the WRA to

> evaluate intra-regional and inter-regional waste movements

> assess the regional development of disposal and recycling

> plan and execute more efficient sampling strategies.

Co-ordinating recycling plans

3.9 The WRA must[39] consider

> what arrangements can reasonably be expected to be made for recycling

> how the plan should provide for these arrangements.

3.10 Waste collection authorities (that is, the districts) have a duty[40] to draw up recycling plans dealing with household waste in their area. However, although recycling is an important option for dealing with household waste, it needs to be set in the context of all the options for dealing with each type of waste.

- Environmentally sound practice for recycling often demands planning that

  - goes beyond district or even county boundaries, and

---

[39] s50(7) of the Environmental Protection Act 1990

[40] s49 of the Environmental Protection Act 1990

- includes other wastes as well as household waste.

The WRA should consider, for each waste type,

> whether the circumstances favour recycling it, and if so
> what the balance should be between recycling and the other options.

3.11 Thus

- the strategy for recycling (and much of the information, for example on waste composition, that is needed to decide it and monitor its implementation) should be derived from waste management planning.

3.12 The WRA should hold regular meetings with recycling staff from each of the constituent waste collection authorities.

Liaison with waste disposal authorities (WDAs)

3.13 Waste management planning should provide sound information on the environmental impacts of the waste management options for all waste that has to be managed in an area but the recommended options will remain as recommendations unless others put them into practice.

3.14 The WDA is responsible for deciding the way household and other waste delivered to it by the districts will be managed; and the way household waste is managed is often a key driver for handling other similar wastes. The WDA will find the WRA's environmental assessment of the waste management options valuable in taking such factors into account in its contracts (see paragraphs 2.34 and 2.35). Thus, waste regulation staff should also hold meetings with officers of the waste disposal authority.

Co-ordination with other bodies

3.15 The WRA should be in constant touch with the National Rivers Authority. In particular the WRA should assess the effects on waste management of the NRA's

> groundwater protection policy
> source protection zones
> estimates of groundwater vulnerability.

Any or all of these may preclude some waste management options for specific wastes in particular areas.

3.16 The WRA will also find it helpful to consult Her Majesty's Inspectorate of Pollution and local environmental health officers. The two groups will between them be able to provide information about

> plant dealing with waste and controlled under Part I of EPA90.

Achieving consistency with development control planning

3.17 As a matter of course (see paragraph 2.23) the WRA should liaise regularly with the planning authorities about

> proposed developments
> likely future planning constraints.

3.18 The WRA should check additionally with the local planning authorities before it designs the wastes survey. They will need the WRA to provide information from the survey: the WRA needs to know what information they want, and in what format.

3.19 Besides liaison at particular milestones, the WRA and the planning authorities should expect to exchange information constantly, and informally. Such exchanges should enhance the quality of development planning and waste management planning alike.

Consultation

3.20 By contrast with co-ordination, consultation is more formalised, takes place at a settled (and often quite late) stage in the process of producing a plan, and is frequently triggered by statutory obligation.

3.21 When a draft waste management plan (or a draft modification of it) is complete, the WRA **must** consult those who might be affected; but additionally **should** consult those who can contribute significant additional data, information or advice.

Obligation to consult other statutory or corporate bodies

3.22 For the WRA's EPA90 duty to consult other statutory and corporate bodies, see box 3.1 below.

Box 3.1 *The waste regulation authority's EPA90 duty to consult statutory and corporate bodies*

---

50 (5) It shall be the duty of the authority-

(a) in preparing the plan and any modification of it, to consult-

  (i) the National Rivers Authority or, in Scotland, any river purification authority any part of whose area is included in the area of the waste regulation authority;

  (ii) the waste collection authorities whose areas are included in the area of the waste regulation authority;

  (iii) in a case where the plan or modification is prepared by a waste regulation authority in Wales, the county council whose area includes that of the authority;

  (iv) in a case where the plan or modification is prepared by a Scottish waste regulation authority other than an islands council, the council of the region in which the area of the authority is included;

  (v) in a case where provisions of the plan or a modification relate to the taking of waste for disposal or treatment into the area of another waste regulation authority, that other authority; and

  (vi) in any case, such persons as the authority considers it appropriate to consult from among persons who in the opinion of the authority are or are likely to be, or are representative of persons who are or are likely to be, engaged by way of trade or business in the disposal or treatment of controlled waste situated in the area of the authority

---

Obligation for public consultation

3.23 For the WRA's EPA90 duty to consult publicly on the draft plan, see box 3.2 below.

Box 3.2 *EPA90 publicity and public consultation for the draft waste management plan*

> 50(5) It shall be the duty of the authority-
>
> ...
>
> (b) before finally determining the content of the plan or modification to take, subject to subsection (6) below, such steps as in the opinion of the authority will-
>
> (i) give adequate publicity in its area to the plan or modification; and
>
> (ii) provide members of the public with opportunities of making representations to the authority about it;
>
> and to consider any representations made by the public and make any change in the plan or modification which the authority considers appropriate.
>
> 50(6) No steps need be taken under subsection (5)(b) above in respect of a modification which in the opinion of the waste regulation authority is such that no person will be prejudiced if those steps are not taken.

3.24 The WRA should undertake these consultations in two phases

> the **first** (pre-consultation) phase early in the drafting of the plan (in case the consultees have information that would materially affect the draft plan)

> the **second** phase on the draft plan itself.

Pre-consultation

On inbound and outbound wastes

3.25 From its survey, the WRA will have initially identified some streams of inbound[41] and outbound wastes. The WRA should consult widely with contiguous WRAs (and those beyond, if they too make demands on waste management facilities in the WRA's area) to get further data about these streams.

With the development planning authorities

3.26 The WRA will reach a point where it believes that, since the information in the draft plan is sufficiently accurate, the draft plan is ready to be circulated more widely. At this point, the WRA should undertake a brief consultation with the development planning authorities

> to ensure that the waste management plan and the development plans are consistent and compatible with each other

> to make the planning authorities aware of the WRA's findings, and

> to ensure they have had a chance to comment.

With other authorities

3.27 The plan should be sent as a **pre**-consultation draft to

> other pollution control authorities (e.g. HMIP), and

> any WRA from whose area the WRA expects to receive wastes.

> the local area offices of the Ministry of Agriculture, Fisheries and Food (MAFF).*

---

[41] 'Inbound' waste is waste brought into the WRA's area; 'outbound' waste is waste taken out of the WRA's area.

* For Wales, throughout this document references to MAFF mean the Welsh Office Agriculture Department.

Penultimate review 3.28 Following these pre-consultations, the WRA should review the draft plan in the light of the comments, and revise it where necessary.

3.29 Where the draft plan allows scope for decision, the WRA should indicate its own preference, and should take account of the pre-consultees' views.

EPA90 consultation 3.30 By this stage the draft plan should contain enough information to provide

- an adequate level of certainty for individual members of the public **and**
- sufficient detail to allow industry to plan its investment.

3.31 The WRA must now

- send the draft plan to the corporate statutory consultees: see box 3.1
- issue the draft plan for public consultation, and
- give the draft plan adequate local publicity: see box 3.2.

Rationale for consultation and publicity 3.32 The WRA's rationale for decisions about consultation and publicity arrangements is to ensure that it

> signals the draft plan's existence: see paragraphs 3.33 to 3.36 below

> allows, for everybody who has an interest (see paragraphs 3.33 and 3.38), enough time and opportunity (see paragraphs 3.39 and 3.40) to consider and comment on the WRA's assessment in the draft plan.

Publicity 3.33 Many people will have a definite and legitimate interest in the draft plan. Consultation will not be effective unless they have learned that they have an opportunity to comment.

3.34 The WRA should normally expect to advertise, in at least one widely circulated local newspaper,

> that the plan is available

> the date by which comments must reach the WRA: see paragraph 3.39 below.

3.35 If other methods of publicising local-authority business in the area are also in customary use – local radio, for example, or computer-driven information points – the consultation on the draft plan should be advertised there too.

Availability 3.36 The public should be able to consult a reference copy of the draft plan

> at the WRA's office, and

> in local libraries, and

> at council offices.

3.37 The public should be able to buy a copy of the draft plan at a reasonable price, from an address clearly shown on the reference copies.

> Copies should preferably be on sale where the reference copies are deposited, but otherwise

> copies should be available by post from the WRA's office.

Automatic issue

3.38 The WRA should send the draft to

> all the bodies consulted during its preparation

> representatives of organisations that had significant involvement in its development: and particularly

> representatives of the local waste management industry

> representatives of industry and commerce as waste producers.

Consultation period and public meetings

3.39 A consultation period of two calendar months should be enough[42].

3.40 If there is substantial public interest in the plan, or a particular part of it, the WRA may wish to hold a public meeting to discuss it. Such meetings should

> be held within the consultation period, and long enough before the end of the period to allow representations after the meeting

> be held outside normal working hours

> be open to all

> concentrate on the plan. Some discussion of existing waste management operations may be unavoidable: it should not dominate the meeting.

Review and determination

3.41 At the end of the public consultation period, the WRA should

> assess the responses

> review and revise the plan as necessary.

Post-consultation revisions

3.42 The WRA normally need **not** consult again on its proposed revisions. The exceptions are

1 if a response may have a material bearing on the function of, or information provided by, a statutory consultee

2 if a revision will materially affect the interests of a statutory consultee.

3.43 If the WRA decides to make such a revision, it should send to the statutory consultee

> the proposed revision

> an explanation of the proposed revision

---

[42] except when the consultation period includes (a) Christmas or New Year or (b) both: the WRA should then add (a) one week or (b) two weeks to the consultation period.

> (if the consultee might be helped by seeing the terms of the consultation response that initiated the revision) a copy of the response.[43]

3.44 The WRA normally need **not** give the proposed revision a wider circulation.

Time limit

3.45 The WRA may reasonably impose a time limit on the statutory consultee. The limit should be conditioned by the nature of the revision.

- Normally, however, the WRA need not allow the statutory consultee more than 3 weeks to respond.

Consent of other waste regulation authorities

3.46 Special arrangements will apply when the WRA consults

> a *sending authority* (that is, a WRA whose area is the source of inbound waste)

> a *receiving authority* (that is, a WRA whose area is the destination for outbound waste from the plan-making authority's area).

3.47 The authority should try to agree with each of these other waste regulation authorities the figures for each type of waste movement across its boundary.

- Under s50(8) the plan-making authority cannot determine the plan (or any modification) if any **receiving** authority withholds its consent[44]

- The WRA, in consulting a receiving authority on the draft plan, should therefore explicitly invite it to include in its response the statement that it consents to the determining of the plan.

The role of the Secretary of State

3.48 Once the responses have been received and, wherever possible, resolved, and, in every case, before the content of the plan or modification is determined, the draft plan must be sent to[45]

> The Secretary of State
> The Department of the Environment
> Room A2.22, Romney House
> 43 Marsham Street
> London SW1P 3PY[46]

so that the Secretary of State may assess its compliance with s50(3).

3.49 When the Secretary of State has assessed the s50 plan's compliance with s50(3), he will let the WRA know. The WRA may then adopt the plan.

3.50 The plan, and modifications to the plan, are public documents.

3.51 The WRA should lodge three copies of the s50 plan with the Department of the Environment **or** the Welsh Office as appropriate.

---

[43] provided the response is not confidential.

[44] unless the Secretary of State agrees that the determination may be made

[45] under s50(9)

[46] The corresponding address for waste regulation in Wales is: Welsh Office, Environment Division 4, Cathays Park, Cardiff, CF1 3NQ

Period of the plan

3.52 Waste management planning is a continuous and often iterative process. The waste management plan is therefore a dynamic document.

3.53 The WRA should **not** seek to provide detailed forecasts for a decade ahead: that is unrealistic. Instead the WRA should concentrate on

> obtaining a sufficiently accurate current picture
> assessing the effects of foreseeable changes.

3.54 This does not mean that the WRA should altogether abandon longer term planning: **the plan period should be 8 to 12 years**.

3.55 The plan should include

1 **a detailed section dealing with the first 4 to 6 years**

2 **a more general treatment of the remainder.**

Why choose these periods?

3.56 The exact timescales for the whole plan and the detailed coverage will depend on circumstances (such as the existence of disposal sites and treatment plants, and the duration of disposal contracts).

3.57 However, it is rare for large-scale incineration projects to be brought to completion in less than 5 years; composting projects, and materials recycling projects, are likely to take 2-3 years to develop. Even when **consented** void space already exists, a landfill may take 2-4 years to develop (though possibly less for sites taking only inert wastes).

3.58 Estimating the mineral void space available for landfill in the future is subject to other uncertainties:

> minerals sites are often successfully restored without the use of controlled wastes, eg by reinstatement to lower ground levels or water recreation areas
> the need for the minerals and therefore the rate of extraction
> policies to increase the use of recycled materials in place of primary aggregates which should lead to a proportionate reduction in extraction.[47]

3.59 Moreover, the consumption of any void space that is available will be affected by sites for land-raising on, for example, poorer quality agricultural land.

3.60 These factors increase uncertainty about both the potential and required void space for landfill. Hence predictions of the available waste management capacity will probably not be accurate beyond 5 years and unreliable beyond a 10-year horizon.

Modifications and revision

3.61 Circumstances will generally change sufficiently that the plan will need to be modified every 2 to 3 years. This interval allows decisions to be made on

> new waste management policies

---

[47] These policies should also, however, decrease the amount of certain wastes (eg construction and demolition wastes) which would otherwise require landfilling.

> new waste facilities

before the end of the detailed planning period.

3.62   Consultation and discussion should be confined to the modification and its effects. Nevertheless, to prevent confusion the WRA should revise and re-publish the whole plan.

Commercially confidential data and level of detail

3.63   Data obtained during a survey may be commercially confidential. The WRA should therefore establish with the supplier whether the data are commercially confidential.

3.64   If data are commercially confidential, the supplier may agree to publication, if the WRA presents the data so as to

> prevent the extraction of any commercially confidential information.

If the WRA cannot do this, it should obtain the data, but treat them in confidence. **The commercial confidentiality of data is not valid grounds to withhold them from the WRA: it is, however, valid grounds to preclude its publication.**

3.65   It is not the function of the waste management plan to duplicate the public register. The WRA should aim to provide a comprehensive picture of waste management without recording fine detail about individual operations.

Chapter 4 The waste management plan

4.1 This chapter sets out the content and format of the waste management plan and explains the calculations and evaluations that the WRA should undertake to develop its strategies for waste management.

Summary of the plan's contents

4.2 Statute[48] specifies the information that every waste management plan must include: see box 4.1 below. In summary, the plan is the WRA's assessment of

> the type, nature and quantity of all waste arising in, being brought into, or taken out of, the WRA area

> the facilities that are available or are needed for recycling, recovery, treatment and disposal

> any special or noteworthy problems in the area

> the alternative strategies and their financial implications, and

> the options that would provide the best overall strategy.

4.3 The plan should take account of

> government policies on waste: e.g. producer responsibility and the 25% recycling target for household waste, and the waste strategy for England and Wales (when published)

> local circumstances and conditions

> the effect of the various options on any targets.

4.4 It may also use information from

> development plans (especially structure plans and waste local plans)

> environmental audits and green charters adopted by local authorities and industry

and will be influenced by

> the sustainability principle

> the proximity and self-sufficiency principles.

The relevance of administrative boundaries

4.5 Neither waste regulation authority nor regional boundaries will restrict the actual movement of wastes.

4.6 Estimates of the requirements for waste management facilities may be based on self sufficiency in the disposal (but not necessarily recovery) of wastes in the region but should be adjusted to take into account other equally suitable facilities near to hand.

---

[48] s50(3) of the Environmental Protection Act 1990; the Waste Management Licensing Regulations 1994

- The proximity principle is breached only if waste is transported a long distance when there is an equally suitable treatment operation or disposal site nearer the point where the waste arises.

Consistent sequence

4.7 Some users of the plan – government, industry, environmental consultants, other waste regulation authorities – may wish to find a particular type of information in more than one plan. This will be particularly so when they are aggregating plans at the regional and national levels.

- Time and effort are therefore saved if waste management plans can be set out in a consistent way.

The guidance in this chapter is given with that objective.

4.8 Paragraphs 4.9 to 4.50 below mention the main sections of the plan, and indicate its scope and content.

Box 4.1   *Information to be included in the EPA90 plan*

---

s50 (3) It shall be the duty of the authority to include in the plan information as to-

(a) the kinds and quantities of controlled waste which the authority expects to be situated in its area during the period specified in the plan;

(b) the kinds and quantities of controlled waste which the authority expects to be brought into or taken for disposal out of its area during that period;

(c) the kinds and quantities of controlled waste which the authority expects to be disposed of within its area during that period;

(d) the methods and the respective priorities for the methods by which in the opinion of the authority controlled waste in its area should be (e) the policy of the authority as respects the discharge of its functions in (e) the policy of the authority as respects the discharge of its functions in relation to licences and any relevant guidance issued by the Secretary of (f) the sites and equipment which persons are providing and which (f) the sites and equipment which persons are providing and which during that period are expected to provide for disposing of controlled (g) the estimated costs of the methods of disposal or treatment provided (g) the estimated costs of the methods of disposal or treatment provided for in the plan;

The foreword

4.9 A waste management plan should normally begin with a foreword. This **briefly** explains

> the statutory obligation to plan

> the broad purpose of the plan

> who produced the plan

> the period covered by the plan

> the organisations and bodies who were consulted on the draft plan

> when the plan was adopted by the authority

> when the plan should be reviewed.

The table of contents

4.10 The table of contents should list the chapter numbers and titles, the main sub-headings, and the maps, figures and photographs. Page numbers should be given. It probably helps the user if paragraph numbers are given as well: but the software may or may not make that easy to do.

The summary

4.11 This should summarise the entire plan (except the foreword) including its findings and recommendations. It should include

> the WRA's waste management strategy, and

> the WRA's main policies for minimising, reusing, recycling, recovering and disposing of waste.

4.12 The aim should be to produce a summary that is comprehensive enough to stand alone and be used as a source of information, yet sufficiently brief to attract busy readers. It should be no more than 4 or 5 pages long.

Chapter 1: the introduction

4.13 This chapter should explain

> the concept and purpose of a waste management plan

> how the WRA has interpreted its plan-making obligations[49]

> how the plan discharges these obligations

> how the waste management plan relates to the area's development plans.

4.14 The chapter should **not** provide a detailed account of the legislation. The text should direct the reader to the legislation, to this paper, to relevant waste management papers, or – if the detail is essential to the plan – to an appendix.

4.15 The chapter should also indicate

> the processes used to prepare the plan, in particular

>> the investigations of waste and waste management facilities

>> the programme of consultation

> the effect of other factors – such as the Waste Strategy for England and Wales – that the WRA takes account of in the plan.

Chapter 2: the background

4.16 This chapter should cover the geographic, demographic and socio-economic features of the area, so far as they are relevant to waste management planning.

a It may draw on

> the area's development plans

> information from the development planning authorities.

b It should mention special features that affect waste production, treatment or disposal: e.g. the presence of major industry.

c It should describe briefly

> administrative arrangements

> present and (estimated) future population distribution

> the main industrial and commercial activities, including agriculture, mineral extraction and tourism.

---

[49] under s50 of the Environmental Protection Act 1990 as amended by the Waste Management Licensing Regulations 1994

    d   It should indicate the main physical features of the local geography, including roads, railways, rivers and canals.

    e   It should describe the area's geology, hydrology and hydrogeology.

- It should relate hydrogeology to the NRA's assessment of groundwater vulnerability.

    f   It should summarise

> existing and planned land use

> designated areas (such as national parks, sites of special scientific interest, areas of outstanding natural beauty) that affect development planning.

Chapter 3: the methods used to obtain data and information

4.17   This chapter should explain the methods used

> to arrive at the waste quantities: for example, the surveys of waste producers and the census of waste management operations

> to estimate the capacities of existing and future waste management facilities.

It should outline the main sources of information, and summarise the way the investigations were carried out.

4.18   A separate volume should be produced with sufficient details of the investigations and methodology to enable others to check and comment on the estimates etc.

> This volume need **not** be published with the plan; but it should be available on request, at a reasonable charge.

Chapter 4: existing conditions

4.19   This chapter brings together the results of the investigations described in the previous chapter. It provides details of

> the wastes arising in, brought into or taken out of the area

> the existing systems of collection, recycling, recovery and disposal, including their capacities.

4.20   For each type of waste (see the list of 26 headings – that is, types – in paragraph 12.3 of this paper) the chapter should state

> the quantity arising, brought into and taken out of the area[50]

> the nature of the waste management facility for which the waste is destined (e.g. landfill, incineration, composting). Where the quantities of a type of waste are significant, that waste type should be dealt with in a separate section.[51]

---

[50]The quantities relating to non-controlled wastes: radioactive, agricultural and mines and quarries wastes will be complete only in so far as their potential impact on controlled waste management is concerned.

[51]For example, waste arising from construction and demolition should be included in this chapter but as a separate section (see paragraphs 12.29-12.43).

4.21 Where practicable, the chapter should include

> details of the waste management industry in the area

> storage, collection and transport handled by the private and public sectors for household, commercial and industrial wastes.

4.22 Information on special waste should appear in this chapter. Special waste should also be dealt with comprehensively in an appendix document: see paragraphs 4.46-4.50 below.

4.23 This chapter should also present more detailed information about the area's different types of waste management facilities, including those for recycling and reclamation. It should indicate (preferably in tabular form)

> the types of site

> the classes of waste permitted

> the maximum capacity in each district or location

> the total current input

> the total capacity available.

4.24 If the investigations have obtained significant facts from industry and commerce about waste reduction, recycling or recovery, these facts should be set out here, so that their effect can be taken into account in the next chapter.

Chapter 5: an evaluation of options

4.25 This chapter describes the options for dealing with the main waste types identified from the survey. Waste Management Paper 1 *A Review of Options* gives guidance on most waste management options; Waste Management Paper 28 gives more detail on recycling. The Planning Policy Guidance note on renewable energy[52] discusses waste-to-energy processes.

4.26 For each waste stream, a range of management options will be possible. Many waste streams include a variety of component wastes. Where these components can realistically be separated, the WRA should consider the management options component by component, rather than merely considering the waste stream as an entity. But if separation is not a practical prospect, the WRA should consider the options for the waste stream as a whole.

4.27 The chapter should discuss the environmental impacts of each option. These need to be consistently, comprehensively and objectively assessed using a method such as life cycle assessment (see paragraphs 4.36 to 4.38).

4.28 This chapter should also compare the costs of the options. The estimates may be fairly broad; but they should allow for the cost of meeting all the statutory requirements that the WRA knows will have effect in the plan period.

---

[52]Annexes to PPG22 (DoE, 1994).

- This includes the long-term requirements, particularly the post-closure controls for landfill.

4.29 The elements of each option should be costed, and these element costs totalled for each option. The costs should be shown in constant prices.

4.30 The total cost of each option should also be shown discounted.[53]

- Since the interpretation of discounted cash flows is sensitive to the choice of discount rate, they should be calculated at two discount rates[54]; the results at both rates should be included.

Chapter 6: the future position, assuming business-as-usual

4.31 In this chapter the plan should assume business-as-usual, and report the quantities of waste (by type, within districts) and the changes over the plan period.

> Business-as-usual implies that waste quantities and composition may change as forecast, but that – for example – policies to promote, or plans for, waste minimisation are **not** implemented, and that recycling and recovery stay at present levels.

4.32 The chapter should

a indicate the useful life of

> existing waste management operations

> operations for which planning consent already exists

b consider other possible waste management operations that are at the application stage (or even earlier)

c show, for each type of waste, when and where any shortfall (or excess) in planned capacity might occur

d identify

> break points in contractual arrangements

> the cessation of operations at the waste management facilities

since

- these provide an opportunity to implement major changes.

Chapter 7: the waste management strategy and its implementation

4.33 This chapter takes the business-as-usual position from chapter 6, and compares its costs and environmental impacts with those of possible options from chapter 5.

4.34 If the business-as-usual position is still the preferred choice, the plan should say so.

4.35 If, however, the WRA considers that the business-as-usual position should be improved, this chapter should review other options that might deliver the improvements.

---

[53]Discounting reflects the cost of money: it has nothing to do with inflation. The combination of discounting and constant prices is correct.

[54]The two rates should be:
> 8% – the present Treasury discount rate and
> 15% – the commercial rate, which is three or four percentage points above base rate.

- The review should take each waste stream (of those streams that affect, or are affected by, the suggested improvement) and assess its impacts and costs.

- The review should also consider existing policies that would affect the amount of waste to be dealt with by the new option.

The role of life cycle assessment

4.36 Determining the impacts of the various options that could be used to manage each waste is a difficult and complex business. Unless all the impacts associated with the particular waste and each option are taken fully into account, then decisions made to improve the environment by managing waste differently may have an adverse overall impact.

4.37 One method[55] that can aid such decisions is life cycle assessment[56]. The DoE has begun a 3 year programme of research into the development of a decision aid to carry out life cycle assessments in waste management. The report of the first stage on developing life cycle inventories for waste management will be published early in 1996[57].

4.38 However, the research will not produce a useable tool for some two years. In the interim, WRAs might find it helpful to draw on the research as it proceeds. Other work in this area may also prove useful, for example, the work by Proctor & Gamble, which includes a spreadsheet programme and data for developing inventories for waste management[58].

Selecting the best strategy

4.39 The basis for the future strategy should be

- the selected range of waste management techniques that produce the lowest overall environmental impact, at an acceptable cost, and with sufficient capacity to meet the forecast quantities of each type of waste.

Integrated strategies

4.40 When all the factors have been taken into account, the strategy will probably, although not necessarily, include several different options for dealing with different wastes, and different components of the same waste at various locations.

4.41 Such integrated strategies are likely to prove the most robust, and to allow flexible responses throughout the plan period.

Chapter 8: conclusions

4.42 This chapter should include analytical descriptions of

> the waste management strategy, including waste management policies

> the waste planning policies

---

[55] Other valid approaches include multi-criteria decision analysis.

[56] Life cycle assessment has the potential to evaluate all the environmental impacts that arise (both directly and indirectly) from an option throughout its entire life cycle (thus taking into account the impacts of transport, depletion of resource such as oil and any change in impacts that might accrue from say replacing raw materials with recycled feedstock).

[57] CWM 128/95: *Developing life cycle inventories for waste management*. Aspinwall & Co., Pira International and the Centre for Environmental Strategy at the University of Surrey, for DoE. Available February 1996.

[58] White, P.R., Franke, M. and Hindle, P. (1995) *Integrated Solid Waste Management – A Lifecycle Inventory*, Blackie Academic and Professional, London.

> the reasons on which the WRA's decisions have been based

> the effect of the policies on the targets in the national waste strategy.

4.43 In addition to the policies to be adopted, the chapter should summarise any other action the WRA proposes to take to implement the plan.

Other topics

4.44 Generally, the waste management plan

> should **only** list the minimum details of the licensed and registered waste management facilities in the area (in an appendix)

> should **not** deal with topics, such as enforcement, that belong in the WRA's annual report.

4.45 However, the plan's appendices

> may usefully explain technical terms (and the like) that are not explained in the body of the plan.

A supplementary volume dealing with special waste

4.46 Under article 6 of the Hazardous Waste Directive, plans for the management of hazardous wastes must be drawn up and publicised.

4.47 The government sees waste regulation authorities providing information to meet this requirement as part of their waste management planning function.

4.48 The user of the waste management plan must be able to distinguish without difficulty

> the sections that deal with special waste

> the quantities of special waste

> the WRA's assessment of methods and sites for dealing with special waste

> policies that bear upon special waste.

4.49 As a matter of convenience the WRA should therefore bring all this together in a supplementary volume.

4.50 The WRA should not need to publish the supplementary volume widely: the information will already have been included in the waste management plan.

Chapter 5 The investigation – background and issues

5.1 This chapter describes

> the general requirements which the investigation should meet

> the background to, and reasoning behind, these requirements.

5.2 The term 'investigation' is used here as it is in the 1990 Act[59]: it goes wider than 'survey'. An investigation may include

> a survey – a study of a group (the population) through a sample chosen so as to have the same characteristics as the population

> a census, where the whole population is studied

> other data gathering and research.

Legal basis for the investigation

5.3 Under EPA90 s50(1) (set out in full in box 2.1 in chapter 2), each waste regulation authority must, in effect

- investigate

    > present and future levels of waste to be dealt with

    > the arrangements available for its treatment or disposal[60]

- assess the best practicable options for dealing with the wastes

- decide the arrangements that ought to be provided to meet that assessment.

5.4 The information to be included in the plan must now[61] include recovery as well as disposal. New objectives[62] apply to waste management plans: see box 5.1.

5.5 The WRA must[63] also from time to time investigate again, in order to decide what changes are needed in the plan, and modify the plan accordingly.

5.6 Thus the waste management plan is a dynamic document. The information on which it is based should be reviewed and updated regularly

---

[59] Under EPA90 s44A (or s44B) – sections inserted by s92 of the Environment Act 1995 – the Secretary of State may direct the Agency (or the Scottish Environment Protection Agency, SEPA) to carry out a 'survey or investigation'.

[60] Treatment or disposal is the term used in EPA90. It includes recovery, recycling and disposal (see paragraph 5.4).

[61] since the WMLR 1994 came into force (on 1 May 1994). Paragraph 9(8) of Schedule 4 to the WMLR modifies s50(3) of EPA90.

[62] added by WMLR Regulations 1994 (paragraph 4(3) of schedule 4); and inserted as schedule 2A to EPA90 by the Environment Act 1995

[63] ss50(1)(d) and 50(1)(e) of EPA90

and often: this information, of course, has uses beyond the document itself.

Box 5.1 *Waste planning objectives: paragraph 4(3) of Schedule 4 to the Waste Management Licensing Regulations 1994*

> [4](3) The following further objectives are relevant objectives in relation to functions under the plan-making provisions-
>
> (a) encouraging the prevention or reduction of waste production and its harmfulness, in particular by-
>
> (i) the development of clean technologies more sparing in their use of natural resources;
>
> (ii) the technical development and marketing of products designed so as to make no contribution or to make the smallest possible contribution, by the nature of their manufacture, use or final disposal, to increasing the amount or harmfulness of waste and pollution hazards; and
>
> (iii) the development of appropriate techniques for the final disposal of dangerous substances contained in waste destined for recovery; and
>
> (b) encouraging-
>
> (i) the recovery of waste by means of recycling, reuse or reclamation or any other process with a view to extracting secondary raw materials; and
>
> (ii) the use of waste as a source of energy.

Producers of hazardous waste

5.7 All producers of hazardous waste[64] are subject[65] to periodic inspection. The frequency will depend on the quantity of hazardous waste stored or produced.[66]

- Where inspections are made annually or more frequently, one inspection each year should be used to investigate all wastes, and identify the scope for minimising their quantity and the hazard they pose.

- Annual updates will provide important information on trends and progress in e.g. waste minimisation.

Special waste consignment notes as data source

5.8 The consignment note[67] contains data on types and quantities of special waste: these data should produce enough information for waste planning purposes.

- Where they do not, the WRA must meet its responsibilities under the 1980 Regulations by ensuring those completing consignment notes provide the information required.

---

[64]Although the precise meaning of hazardous waste depends upon the interpretation of the Directive, **for waste management planning purposes** it is reasonable to assume that it will eventually apply to all those wastes that in the UK are, or will be, regarded as special.

[65]Hazardous Waste Directive (91/689/EEC), article 2.3.

[66]Where the quantities are such that a licence is required, the frequency of inspection should be on the basis of Waste Management Paper 4 (3 times per year); other significant producers of special waste should be inspected annually; and for very small producers (such as those on some collection rounds) the frequency will be less – perhaps once every 3 to 5 years (unless there are indications that a higher frequency is required in particular cases).

[67]under the Control of Pollution (Special Waste) Regulations 1980: SI 1980 No. 1709

Frequency and nature of the investigation

5.9 If the WRA is to make informed decisions on waste management from day to day (and certainly if it is to recommend sensible arrangements for dealing with waste in the future), it needs to know about present patterns and trends of waste movement and management in its area: a discrete survey every five or ten years **cannot** provide this knowledge.

- The WRA must be able to relate information from one year to other years, and from one locality to other localities. The investigation must be structured so as to make this possible.

5.10 The methods of investigation will depend on

> the nature of the information required

> the resources available

> the number and type of different sources.

5.11 In general, however, the investigation should use

> surveys – i.e. statistical sampling – of waste producers

> a census – i.e. a 100% sample – of existing waste facilities

> a study of waste movements and arisings

> a review of the available options for dealing with waste

> estimates of the cost of each option.

5.12 The WRA should **annually** investigate in its area the

> waste arisings

> waste movements

> waste management operations.

- Waste arisings should, however, be subject to a more extensive survey in Year 1 and every 3 years thereafter (see paragraph 7.7).

Regional sampling

5.13 Waste arisings surveys should be planned across the region as a whole. This will mean surveys will be more

a efficient

b accurate

c consistent.

Reporting year

5.14 The reporting period should be the financial year, 1 April to 31 March. Data should however be gathered in quarters, so that seasonal variations may be detected and information for calendar years calculated if required.

5.15 The connections between the data **items** (such as waste arisings, in 5.12), the data **sources and methods** (such as the census of existing waste facilities, in 5.11) and the report **topics** (such as 'existing and planned sites') are discussed at paragraphs 5.19 to 5.22 below.

Scope of the investigation

Both origins and destinations

5.16 No survey or investigation is adequate unless it enables the WRA to make reasonably up-to-date and accurate estimates of all the significant types of waste. The estimates should be made

> at the places where the wastes are produced, **and**

> at waste management facilities.

5.17 The WRA should **not** rely solely on **either** the input figures supplied by licence holders **or** the process-related information obtained from industry (and other producers of waste) in the survey of arisings. For its estimates of the wastes to be dealt with, it should get data from

> waste producers **and**

> waste management facilities.

These **must** include data from

> waste management operations that are exempt from licensing.

Although not licensed, such operations still deal with controlled waste and when taken together may account for substantial quantities.

Usefulness for cross-checking

5.18 Both sources – waste producers and waste management operators – have their merits: both should be used to cross-check and confirm the data.

> The survey of waste producers should yield information on waste streams from different processes and operations. It incidentally allows the WRA to provide advice on waste minimisation.

> Information from waste management facilities should provide data of greater accuracy on waste quantities. From these data the WRA should be able to cross-check the data on arisings.

> Investigation of waste inputs at waste management facilities should also enable the WRA to cover a higher proportion of waste movements in each annual survey.

Data items paired with data sources

5.19 In short, the WRA's investigation gathers three main kinds of data item, each paired with one main source. The three pairs are

- **waste arisings** – data mainly from the **survey** of industrial and commercial waste producers

- **waste movements** – data mainly from **site returns** from waste management operations, but supplemented by on-site inspection of wastes and *duty of care* documents[68]

- **waste management operations** – data (details of throughputs) from an **annual census** of the operations.

Topics

5.20 However, the investigation's topics (models, in effect) should include

---

[68]There is a risk of double-counting some wastes from waste management facilities. A consistent treatment to avoid this is explained at paragraphs 10.16 and 10.17.

    a   present and future amounts of wastes

    b   existing and planned sites

    c   the environmental effects (beneficial as well as damaging) of the methods of disposal, recovery and treatment

    d   the costs of the methods of disposal, recovery and treatment.

5.21   Some of these topics depend for their data on data items (see 5.19) of more than one kind, and thus from more than one source.

> For example, **b** (existing and planned sites) depends on data about arisings **and** movements **and** operations.

5.22   The topics also depend on each other for information, and therefore act upon each other. For example

> **d** (costs) depends on **c** (environmental effects)

> **a** (present and future amounts) depends on **d**

> **b** (existing and planned sites) depends on **a**, **c** and **d**.

5.23   Only multiple-source datasets will provide the information the WRA wants.

Multiple sources also minimise survey demands at annual updates

5.24   If the WRA uses multiple data sources, it should be able to update the investigation annually without necessarily needing a full survey.

- Each year's investigation should collect a **full** set of data on waste movements, waste facilities and major producers of **special** waste.

- But the WRA (with other WRAs in the region) should structure its survey of **other** waste producers so that a smaller sample is enough in the intervening years. This smaller sample should

    > allow quantities to be cross-checked

    > provide information on waste production trends in different industries.

Costs and throughputs: confidentiality of data

5.25   Some information on costs and throughputs may be commercially confidential: see the discussion of confidentiality in chapter 3, paragraphs 3.63 to 3.65 above.

5.26   If the WRA acquires commercially confidential information, the WRA must itself treat it as confidential.

- For example, the WRA must not publish (or otherwise disclose) the information[69] in a form that would enable a competitor to identify the unit cost of the disposal or treatment of an individual waste stream for a specific customer at a particular site.

5.27   Given the existence of these safeguards, industry should co-operate freely in the survey.

---

[69] Except with the consent of those concerned

Other sources of information

5.28 Information may also be found

> through local knowledge, including that of waste collection authorities

> in the public pollution control registers – that is, the local authority air pollution control register and the Integrated Pollution Control (IPC) register: and more particularly

> in applications for IPC authorisations, which, in some cases, contain information about the wastes produced.

Non-controlled wastes

5.29 The WRA's obligation[70] is to provide information about **controlled** wastes. But to estimate how much capacity each waste management option demands, the WRA needs to take into account the capacity likely to be absorbed by **non**-controlled wastes: see chapter 12, paragraphs 12.61 to 12.75.

5.30 In particular

> spreading animal slurries from agricultural premises uses land that might have been suitable for spreading some kinds of industrial waste;

> mineral extraction wastes backfilled into the void need to be quantified because they may consume space that might otherwise have been assumed to be available for controlled waste; and

> where a site will be restored using biodegradable wastes, and minerals waste suitable for intermediate/daily cover (e.g. overburden or dust from crushing operations) is available, then this may mean the site is not an outlet for controlled wastes often used as cover – eg construction and demolition wastes such as engineering spoil.

Waste arisings – weights and volumes

5.31 For accurate estimates, weighing waste is the only wholly satisfactory method. Weighing, by quantifying waste and therefore its real costs more accurately, also encourages the waste producer to take more interest in minimising wastes. Weighing waste is discussed in more detail in paragraphs 6.36 to 6.37.

Area of investigation

5.32 WRAs should co-operate – nationally to plan surveys and regionally to carry them out: but the data obtained should enable information to be related to local authority areas both for production and recovery/disposal. This breakdown applies especially to the study of waste movements: WRAs should obtain information for each type of waste deposited in, arising in, taken out of, or brought into, each local authority area.

5.33 Therefore, if the WRA ensures in its investigations that wastes are always referenced to the local authority of origin and deposit, it will be able to identify waste flows more accurately; and inform waste disposal, local planning and waste collection authorities.

- This will inform the development of recycling plans and provide information on the arrangements for the collection, recovery and disposal of household and commercial wastes.

---

[70]under both the Waste Framework Directive and s50 of EPA90

Who should carry out the investigation?

5.34   The WRA is responsible for deciding how much waste will have to be dealt with. The WRA should not abdicate this responsibility to another body.

5.35   Local planning authorities in particular will wish to use the figures derived from the investigations carried out by WRAs. The WRA should provide the results from individual investigations to planning authorities and to regional planning groups as required.

Staffing of surveys

Qualified staff

5.36   To plan and carry out the investigation the WRA should use only appropriately qualified staff. Planning the survey and interpreting its data will require statistical expertise or advice. Those who visit industry should be capable of advising industry on waste management, and particularly on waste minimisation, recycling and recovery. This will

> minimise the number of separate approaches to industry

> fulfil statutory obligations

> produce results in which industry can be sufficiently confident.

Limited role for casual and temporary staff

5.37   Casual or temporary staff have in the past assisted in investigations of waste arisings, particularly in collating or encoding data. Because waste management planning now needs better data, casual and temporary staff should **not** be relied upon unless they have adequate

> experience

> training

> supervision.

5.38   Even then, casual and temporary staff will probably have a minor role in the investigation: they cannot credibly advise industry about waste management and waste minimisation.

Chapter 6 Survey of waste arisings – estimates and their improvement

Introduction

6.1 Research carried out on behalf of the Department of the Environment in 1990[71] concluded that four elements are essential in estimating industrial waste arisings. The same elements will thus be common to any properly designed sampling scheme for estimating waste arisings. They are

1. **identification** – a means of identifying the population (i.e. all the firms in the area) to be surveyed: see paragraphs 6.3 to 6.13 below

2. **dependent variables**[72] – a piece of data for each of the firms identified that is related to the amount of waste likely to be produced.

3. **classification** – a means of dividing the population into groups where the waste producing activities of its members is broadly similar; classification is considered in paragraphs 6.18 to 6.19 below.

The survey is used to produce a fourth element

4. **waste factors**[73] – the factor that relates waste produced to the dependent variable.

6.2 The chapter then considers

> how estimates may be improved (paragraphs 6.30 to 6.40), and

> work put in hand by DoE for this purpose (paragraphs 6.41 to 6.59).

Identification

6.3 The chief requirement for an identification procedure is that, for the given waste, it identifies **all** the waste production activities in the area (i.e. the entire population from which the sample should be drawn).

> For some types of waste, e.g. clinical wastes, the sources will mostly be obvious and readily identifiable.

> For others, e.g. fragmentiser residues, the WRA should be able to obtain a complete picture from experience and local knowledge.

> For others again, the WRA may need help from another agency: in calculating agricultural waste production, for example, the local MAFF office should be asked to help identify all farms in the area.

---

[71] CWM 022/90: *Commentary on techniques for determining arisings of industrial solid wastes.* Aspinwall & Co., for DoE. Available from the Waste Management Information Bureau.

[72] Dependent variables in particular are considered at paragraphs 6.14 to 6.17.

[73] See paragraphs 6.27-6.29 below.

Identifying industrial and commercial premises

Sources

6.4 Probably the most difficult job is to identify all industrial and commercial waste producers: this is now considered.

6.5 The WRA aims to use for survey purposes a list of firms that completely represents the population.

6.6 Sources for the data are reviewed in CWM 022/90. They include

> the Census of Employment

> the Census of Production

> the Business Database (part of the Yellow Pages)[74]

> local chambers of industry and commerce

> marketing information companies.

6.7 The Census of Employment and the Business Database have the most extensive coverage and the use of one or other is recommended. Their advantages and disadvantages are summarised in table 6.1 below. Other databases may provide a useful cross check on survey information.

Table 6.1

Comparison of the Census of Employment with the Business Database for use as a sampling frame

Criterion	Census of Employment	Business Database (Yellow Pages)
Does it include company name and address?	Yes	Yes
Does it include the postcode?	Yes	Yes
Does it include the telephone number?	No	Yes
Does it state the number of employees?	Yes	Yes and no (Employee bands only)
Does it include the SIC code?	Yes	Yes
Does it identify the local authority district?	Yes	Yes
Is it free?	Yes (in most areas)	No
Is it easily available?	Yes and no (depends on the area)	Yes
Is it available on disk?	Yes	Yes
How comprehensive is the coverage of firms?	Good on large firms; but, in most years, only covers a sample of those employing less than 25 staff	Better than Census of Employment for small firms; but only those firms who pay for an entry in the Yellow Pages are included
How up to date is it?	Usually 2 years or more out of date	No more than 1 year out of date
Other comments	Some of the SIC codes are inappropriate to the actual business activity	Some of the SIC codes are inappropriate to the actual business activity

6.8 The Census of Employment is a biennial survey conducted by the Department of Employment. Information from the Census is generally available two years after the survey was carried out. In most years, a sample is taken of firms employing less than 25 staff so the listing obtained will not include every small firm in the area. The 1993 Census (available mid-1995) was one exception and includes every company. The 1993 Census is also the first to report in SIC(92) order.[75]

---

[74] The Business Database may be contacted on 01753 583311

[75] It should be noted that the self-employed, members of the armed forces, private domestic servants and those on a government training scheme without a contract of employment are not covered by the Census. Full time and part time employees are separately listed, part time being defined as those who work 30 hours or less each week (40 hours for agriculture and horticulture).

6.9 The local planning authority holds Census of Employment data for the area. The WRA can get details of commercial and industrial waste producers from the planning authority[76].

6.10 The Business Database will supply a random sample of premises by SIC number.

6.11 These census data should be supplemented with, and cross-referenced to, other sources of information, so that the WRA obtains details of as many business premises as possible: see paragraph 6.3 above.

> Many local authorities hold their own database of businesses, which may provide useful information.

Data in the WRA's list

6.12 The WRA's list should include

1. name of the company
2. premises address (the waste producing site is not necessarily the head office)
3. postcode
4. Standard Industrial Classification (SIC) code, 1992 version: for the SIC, see paragraphs 6.20 to 6.23 below
5. numbers of full-time and part-time employees
6. local authority (district).

Electronic list management

6.13 The list should be held on an electronic spreadsheet or database to make manipulation, analysis, sorting and mailing easier.

Dependent variable

6.14 A listing of industry that contains only names, addresses and the SIC code will be of little value in estimating waste arisings. The list must contain either data on waste arisings or at least one item of data for each firm that can be related to waste arisings.

Characteristics of the dependent variable

6.15 The dependent variable has two properties.

1. The variable correlates satisfactorily with waste arisings in the set of industry the firm belongs to.
2. The variable is readily available for each firm in the set.

> For agricultural premises, this variable will be the area of crop grown or numbers of animals reared.

> For industry and commerce the best, widely available variable is number of employees.

• To estimate waste arisings for industry and commerce, the WRA should use number of employees as the dependent variable.

6.16 For much of the minerals extraction industry, the amount of waste produced will depend on several factors including the depth of

---

[76] DoE is aware that some WRAs have been refused access to the Census of Employment data. Paragraph 2 of Schedule 22 to the Environment Act 1995 gives the Environment Agency direct access to these data.

overburden, the thickness of the mineral layer and the amount of unwanted material within the mineral layer. Thus, general correlations between the quantity of waste and the number of employees or the amount of mineral produced are not usually possible. The WRA should seek information from the minerals planning department. Any uncertainties that remain can be resolved by contacting the companies themselves.

Sources for values for the dependent variable

6.17   Many of the values for filling in the dependent variable will derive directly from population or classification data: the sources of these data will usually be well known.

> Thus, for agricultural wastes, the WRA would get the dependent variables factors – such as acres under each type of crop, or numbers of livestock – from the MAFF local office.

> For industry and commerce, the number of employees for each firm is given in the Census of Employment.

Classification

6.18   While a listing of industry and commerce including a dependent variable is an important step forward, the better the correlation between waste arisings and number of employees, the more accurate will be any estimates produced. The WRA's objective is to subdivide its list of firms until the list consists of activity groupings: the characteristic of each grouping, called in this paper a *set*[77], is that every firm in the set produces broadly similar types of waste.

6.19   Industries using similar processes will obviously produce similar types of waste. Thus, to improve estimates, businesses of similar types need to be classified together. The most convenient and comprehensive classification system is the Standard Industrial Classification (1992)[78]: the following paragraphs describe the aspects of SIC that are significant for this paper.

Standard Industrial Classification

6.20   The Standard Industrial Classification (SIC) lists categories of business, grouping them in similar types of activity. Table 6.2 illustrates the hierarchy of categories.

6.21   For the purposes of the survey, the WRA should use the 1992 version of SIC.

6.22   Waste arisings need to be estimated by industry type. The list used to select firms must therefore include the SIC code for each firm: see paragraphs 6.5–6.13 for the contents of the selection list.

Sort at SIC class or sub-class level

6.23   At the Division and Group levels, as table 6.2 suggests, many of the SIC headings are too broad to allow generalisations about waste production across the division or group.

> For another example see SIC(92) Division 50 – sale, maintenance and repair of motor vehicles and motor cycles (and retail sale of automotive fuel). The waste produced by selling cars will be entirely different from that produced by their repair.

---

[77] 'Class' or 'group' might have been more natural, but each already labels one of the levels in SIC.

[78] Central Statistical Office, (1992), Standard industrial classification of economic activities 1992. HMSO, London. ISBN 0 11 62550-4

6.24 When compiling the list from which the sample is to be selected, firms should therefore be sorted at the most detailed level available: that is, SIC class or sub-class.

Table 6.2
Illustrative section of SIC 1992

Section	Sub-section	Division	Group	Class & Sub-class	Description
D MANUFACTURING					
	DA MANUFACTURE OF FOOD PRODUCTS, BEVERAGES AND TOBACCO				
		15 MANUFACTURE OF FOOD PRODUCTS AND BEVERAGES			
			15.1 Production, processing and preserving of meat and meat products		
				15.11	Production and preserving of meat
				15.11/1	Slaughtering of animals other than poultry and rabbits
				15.11/2	Animal by-product processing
				15.11/3	Fellmongery

Correlating with previous SIC-based data

6.25 The WRA may have existing survey records that use the SIC(80) codes, and may wish to

> compare aspects of those surveys with this one, or

> establish sample sizes.

If so, the WRA should use the CSO's correlation matrix[79] to ensure the results are as consistent as possible.

6.26 However

- wholesale conversion of data that use SIC(80) codes is not recommended: there are few one-to-one correlations.

Waste factors and estimation

6.27 The sample survey for each set of the population should be used to produce an estimate for the amount of each type of waste produced per employee for the set – from here called the *waste factor*.

6.28 The dependent variables in each set should be summed (e.g. the total number of employees in an SIC class). This sum is multiplied by the waste factor to give the grossed-up total of the area's waste from that set.

6.29 A hypothetical example follows.

> The WRA defines one set as the *retail sale of floor coverings* (SIC code G 52.48/1)

> The number of employees is the **dependent variable**. The Census of Employment figures show there are 687 people employed in **the WRA's area** in this class of activity at 138 locations[80].

---

[79] The matrix can be had in hard copy and as an electronic spreadsheet. Central Statistical Office. Business Monitor PO1009: *Standard Industrial Classification of Economic Activities: Correlation between SIC(92) and SIC(80)*. London: HMSO: 1993. ISBN 0 11 536311 4.

[80] The WRA's database design will probably need to include a file of set totals: in fragmented industries the WRA may well **not** have values for the dependent variable in the individual firm's record.

> The sample survey for **the region** includes 41 shops in this SIC sub-class employing a total of 211 people. For one type of waste, rubber based underlay, the **waste factor** (the annual quantity of waste per employee) for this class is calculated[81] as 8.43 tonnes/employee[82].

> The annual total of waste rubber underlay from carpet retail in the WRA's area would be 687 × 8.43 tonnes per year: that is, 5,790 tonnes.

Improving the estimates

6.30 There are three ways of improving estimates of waste arisings:

1. recording the weight (rather than the volume) of waste arisings
2. estimating more accurately the waste produced by the largest producers
3. improving the correlation between the number of employees and waste arisings.

The importance of recording the weight

Weighing for the survey

6.31 When the WRA surveys waste arisings at producers' sites, the quantities of each waste must be measured with minimal error: consistency and accuracy of the survey data are otherwise jeopardised.

6.32 Waste can be most accurately described and most easily weighed separately at the point where it is produced. Very little waste at present is so described and weighed.

- The WRA should seek the co-operation of as many waste producers as is practicable.

- The co-operating producers should weigh their wastes. Wherever possible they should weigh them **before** any mixing takes place.

    > The producers should pass these results to the WRA to update the data – locally, regionally and nationally.

6.33 Waste should be weighed over a 2 or 3 month period. The period should be longer if the production process is seasonal or where there is a large variation in the quantity of each load.

6.34 If the waste has not been weighed at the producer's premises, it should if possible be weighed when it arrives at the waste management facility or en route at a public weighbridge.

> If weighing is not possible, the volume of waste should be estimated, and standard conversion factors used where available[83].

6.35 When waste is being weighed, the WRA may find it helpful at the same time to estimate the volumes of the wastes being moved. The weight-to-volume ratios from these extended surveys can then be used to improve the conversion factors mentioned in paragraph 6.34 above.

---

[81] This is not likely to be the arithmetic mean but the minimum variance unbiased estimate of the mean (MVUE).

[82] This, like the rest of the example, is **hypothetical**.

[83] DoE will provide standard conversion factors in the development of the national waste database.

General benefit of weighing

6.36   All waste producers should be encouraged to weigh their waste. They themselves will benefit: they will get a more accurate view of the costs they incur in managing waste, and of the resources they lose in the waste they produce.

6.37   If a waste producer cannot weigh all waste on site, he should be encouraged to weigh a sample of loads and record their volumes, so that the volumes can be linked to the weights.[84]

Concentrating on larger producers

6.38   The WRA may reasonably assume that

> 80% of the waste will come from 20% of the waste producers

> more reliable and cost-effective *overall* estimates will be produced by accurately estimating the waste from this 20% of producers.

6.39   If a firm is in this group, the WRA should obtain weights for all the firm's waste streams.

> When the WRA first approaches these firms, it should explain why it needs weight data, and should suggest that the firm might contact its disposal contractor for these data if it does not have them already.

Better correlation

6.40   The correlation between waste arisings and employee numbers (as described in paragraph 7.20) may be improved by narrowing the industrial grouping over which the estimate is to be made; but this may mean that the total number of firms sampled has to be much larger.

Action to resolve difficulties

6.41   DoE has commissioned research to resolve some of these difficulties.

The national waste database

6.42   The national waste database will be held by the Environment Agency and will initially be based on data from sample surveys funded by part of this research. The information available will include

> measurements of arisings for various wastes from different classes of industry and commerce

> correlations between measured amounts and numbers of employees

> the results of the national household waste analysis programme (NHWAP): see paragraphs 12.10 to 12.12 below.

6.43   The national waste database will be accessible for

> WRAs, who can use the database to strengthen their survey work

> firms, who can use it to improve waste minimisation by comparing their waste production performance with the range and mean[85] for their industry.

---

[84] It would be helpful, as paragraph 6.35 suggests, to record the volumes of waste in any case, even where all waste is weighed, to enable more accurate conversion figures to be derived for wastes where weighing of any sort is not feasible.

[85] Mean here means, as before, the minimum variance unbiased estimate of the mean (MVUE).

Maintenance of the database

6.44 WRAs, and others using the database, will be asked to improve it by contributing their own information. This will increase the coverage and the accuracy of the estimates that can be made using the database.

6.45 All information supplied will be validated to ensure its currency and accuracy. Data on industrial and commercial wastes will be separated into

> those reported as actual weights

> those converted from volume figures[86] using the standard conversion factors[87].

National waste exchange

6.46 DoE intends that the information stored on the database should incorporate waste quantities and waste composition. The database will thus provide source data for a national waste exchange.

National system for waste classification

A uniform approach essential

6.47 WRAs need to co-operate, within regions and between regions, in planning and carrying out the survey. Such co-operation is essential if WRAs are to

> identify waste movements between WRA areas

> determine the availability of waste management facilities in other areas

> achieve waste management planning's full potential to provide the essential information at local, regional and national levels.

6.48 Even with full exchanges of information and a sound statistical approach, the ability to aggregate data regionally and nationally will depend on uniform description and classification.

Uses of waste classification

6.49 No two loads of waste are **absolutely** identical in their composition. But for compactness and comprehensibility, the WRA should be able to group together the wastes that have similar characteristics.

> Such groupings may be by source, or by general properties, or by composition.

6.50 All WRAs use some means of waste classification. Many use more than one, since few use the same classification for waste management planning as for waste licensing. Yet waste management planning depends on the licensing system: waste management planners need (for instance) to cross-check waste producers' data against waste management industry data: these industry data take their form from conditions in waste management licences.

- Discontinuity in classifications between the waste licensing system and the waste management planning system thus, at best, causes extra work; but, worse, it may remove valuable information because the data cannot in practice be used.

---

[86] DoE hopes that the volume-derived figures will gradually be superseded as more information on weights of waste becomes available.

[87] The production of standard conversion factors will be undertaken as part of the research to produce the national waste classification scheme and the national waste database.

6.51 All investigations necessarily involve some assessment of inter-authority waste movements. These movement data need to be checked. The results of individual investigations (including their raw data) need to be collated and aggregated at regional and national levels. If all this is to be done, the same waste must be described the same way, wherever the description originates and whoever writes it.

6.52 Yet further: although European policy emphasises that Member States should achieve self-sufficiency in waste disposal, and increasingly limits the trans-boundary movement of wastes, the European legislation that aims at these objectives relies on classification of waste. Hence there are strong grounds for classifying wastes consistently, both nationally and throughout Europe.

The European Waste Catalogue and DoE codes

6.53 The European Waste Catalogue (EWC) and the DoE waste codes unfortunately will not serve the purpose.

> The EWC lists mainly industrial wastes. It does not deal adequately with other wastes, or with the chemical composition of wastes.

> The DoE waste codes adequately reflect the chemical composition of the waste, but do not describe its source.

Development of a national waste classification system

6.54 To achieve consistency of classification, the Department of the Environment has funded the development of a national waste classification system.

• A consultation draft will be issued in December 1995

6.55 The system has been designed to incorporate the substance of every waste classification system currently in use by WRAs. It should provide for information on

> the composition of the waste

> the source of the waste.

6.56 The government intends to introduce the national waste classification scheme on 1 April 1996.

The importance of the waste production process

6.57 There are strong indications that estimates of industrial waste production can be made more accurate by correlating waste quantity with production process.

• The new national waste classification system will allow for identification of the industrial process that has produced the waste.

6.58 WRAs can contribute to the analysis by identifying in their surveys the process that produces the waste.

> As well as leading to more accurate estimates, this will assist the WRA to identify opportunities for minimisation, reuse and recycling.

Wider, integrated use

6.59 The national waste classification system is being developed for industry as well as for WRAs.

> **Industry** should use it to describe waste on duty of care transfer notes.

> **Holders of waste management licences** should use it in classifying each load of waste, and in making returns to the waste regulation authority.

> **WRAs** should use it in waste management licensing and in waste management planning.

Chapter 7 The survey – planning, pitfalls and presentation

7.1 This chapter

> gives guidance on planning the survey of industrial and commercial waste producers

> identifies and explores some of the key issues.

Definitions

7.2 Here, as elsewhere, the paper uses 'data' to mean the raw data item collected by the WRA's field staff: e.g. in March 1995 Brown's Engineering Works sent for disposal at Jones' recovery plant 50kg of used oil.

7.3 To produce 'information' – which is the objective – the WRA might aggregate all the stocks and flows of waste oil from all the waste producers for the year 1995: cumulating with later years, the WRA would be able to construct a trend line for oil recovery and final disposal.

Planning the survey

7.4 Thorough and careful planning is essential to the success of the survey. Planning will ensure that the WRA collects only data that are needed, and that it does so accurately and efficiently.

Regional co-operation and approaches to the survey

7.5 Waste-planning data and information will need to be capable of aggregation

> within regions, and

> from regions to national level.

All WRAs should therefore gather, record and store data in compatible formats.

7.6 A more immediate reason for adopting compatible formats is to allow the use of multi-authority datasets – when limitations on the range of industry for which waste factors can be used mean the number of samples required to produce information at the required level of detail or accuracy cannot be resourced from a single authority.

> This is particularly likely to happen when narrowing down waste factors to single types of business: the number of premises able to be sampled in any one WRA area may not allow the required estimate to be made.

> Where it is possible to produce an estimate, the sampling errors may be so large, or the confidence limits so low, that the estimate is of little or no value for waste planning.

Hence **WRAs should co-operate to plan regional surveys.**

7.7 WRAs should also use the national waste database as an essential part of waste management planning. It will provide information on the mean unit waste arisings for classes of industry that are not being sampled in an area.

Objectives of the survey

7.8   The aim of the survey is to collect data from which the WRA can develop a sound information base for waste management planning. However, the **objectives related to individual waste components** will influence the choice of sampling protocol, the cost of the survey, and the staff time spent on it.

7.9   The objectives should be clear and precise. They should be written down so that they can be consulted throughout the survey[88].

7.10   The survey of itself can produce only data: the WRA's objective is to turn those data into information. The objectives should define the information the survey is to produce, and the data needed to produce it.

- Throughout the survey process, and particularly when designing questionnaires, the WRA's staff should keep the information needs in mind; otherwise they risk omitting to collect essential data.

7.11   Table 7.1 describes the minimum information (outputs) for which data should be collected, together with the reasons. If more detailed figures are required, or different issues are to be covered, they should be defined at this stage.

> Thus, if the WRA wishes to assess the effects of waste minimisation advice at the area's engineering works, it will need to know – for example – the reduction in oil sent for disposal or treatment. The sampling strategy must take this into account.

---

[88] C A Mosser and G Kalton in their *Survey Methods in Social Investigation* say 'Not the least important purpose of such a statement is that it will clarify the surveyor's own mind and thus probably lead to a more efficient enquiry (and one that will be more easily explained to respondents). Failure to think out the objectives of a survey fully and precisely must inevitably undermine its ultimate value; no amount of manipulation of the final data can overcome the resultant defects.' (p 44)

Table 7.1
Data gathering objectives and rationale

Data to be collected	Rationale
Type and quantity of waste arising	
Quantity and composition of waste (including packaging waste) by industry group of origin	• Assessing the need for new waste management facilities of each type according to the industry's projected growth • Assessing which are the largest waste-generating industry sectors so that the WRA can tailor its waste-minimisation advice to fit them • Assessing whether the industry could increase its use of non-disposal options • Providing base data on the types and quantities of all waste, including packaging waste [89] • Assessing all waste, including the effect the diversion of packaging waste from the waste stream might have on requirements for new waste management facilities
Geographical origin	
Quantity and composition of waste by administrative district and industry group of origin	• To inform planners and regulators at district level about the types and quantities of waste being generated in their area • To help land-use planners in optimising the location of waste management facilities according to the proximity principle
Treatment or disposal method	
Quantity and composition of waste by method of treatment or disposal and industry group of origin	• Quantifying the current use of non-disposal options • Evaluating the possible local impact of government policy on particular disposal options, and on the need for new sites and new types of site • Assessing the potential for increasing the use of non-disposal options • Helping district recycling planners to choose which waste materials to focus on • Assessing how changes in the profile of industry affect the need for particular kinds of waste management practice and of waste management facility
Destination of waste	
Quantity and composition of waste by district and county of destination	• Quantifying the waste being exported so that the WRA can measure progress in implementing the proximity principle and the regional self-sufficiency principle

Incorporating other visits into the survey

7.12 The WRA will only get meaningful results if it selects its sample of firms solely on statistical considerations.

> The WRA may combine the survey with its obligatory inspections of hazardous waste production. Unless the sampling strategy includes all the producers of hazardous waste, the WRA's staff must ensure that they include in the information used to produce

---

[89] The adoption of the EC Directive on Packaging and Packaging Waste will affect the diversion of packaging waste away from disposal. Information on the arisings and fate of packaging waste is likely to become a key issue: gathering information on the quantity and nature of packaging waste should therefore be one of the data objectives. The precise nature of the information has not yet been finalised but is likely to require both material types and uses. DTI leads the negotiations in Europe on this Directive and will be able to advise on the information required.

estimates from the survey **only the data from those producers selected for inclusion in the sample.**

7.13   Chapter 13 deals with sample selection in more detail.

Accuracy and sample size

7.14   It is beyond the scope of this guidance to cover all the statistical theory that might be required to carry out a survey and, particularly here, to determine the size of the sample in each case. However, unless the survey is planned, and samples are selected, in a statistically sound manner, it is likely to negate the results of the survey. Hence,

- the survey should not be planned or undertaken in the absence of suitable statistical advice or expertise.

Nonetheless, statistical requirements are so fundamental to the survey that an outline of some of the basic principles is given here.

7.15   The accuracy of an estimate produced from a sample of a population is independent of the size of the population being sampled.[90] For a given level of confidence[91], the size of the sample needed is a function of the *variance* of the population (the degree to which the members differ from the mean) and the accuracy of the estimate required.

- Thus a sample from 20,000 firms in an SIC class will produce estimates as accurate as the same size sample from 800 firms.

7.16   **Waste factors**[92] for industry and commerce become more precise the narrower the range of waste producing activities over which they are considered. Where high levels of accuracy are required in estimates, the WRA is obliged to sample at the most detailed SIC category available (class and sub-class). The statistical accuracy[93] of each of the separate waste factors will be determined, as above, by the size of the sample (although, in practice, samples in excess of 10% of the population are unlikely to produce any significant increase in statistical accuracy).

Required accuracy determined nationally

7.17   WRAs should decide nationally what level of accuracy is required for each information item to be derived from the survey and the level of confidence to be associated with each one. WRAs will wish to take into account other probable sources of error and their magnitude:

> the method of measurement

> the means of estimation, and

> mistakes in reporting etc.

This will enable waste regulation staff to establish a realistic statistical accuracy.

---

[90]However, as with many applications, the variability of waste production per employee increases with the range of industries included in the population being studied – and, consequently, the size of the population being sampled.

[91]For normally distributed populations (and certain other distributions), the level of confidence, also known as the confidence interval, is the probability that the mean value for the population lies within a given range calculated from the sample.

[92]waste production per employee

[93]Here, statistical accuracy is another way of describing sampling error – the error in the estimate due to studying only a sample of the population.

Calculating the required sample size

7.18   Once WRAs have decided the accuracy and the level of confidence, the minimum sample size for each of the SIC classes (or size bands within classes) should be calculated. Where information (e.g. the standard deviation) about the population exists, this can be a relatively simple procedure[94]. However, this is unlikely to be the case.

7.19   For normally (or near-normally) distributed populations, it is still possible to calculate the minimum sample size required, but information from a previous survey, or a pilot survey, will be necessary to calculate the sample standard deviation. This can then be used to together with *Student's t*[95] distribution tables to calculate the minimum sample size required.

Waste production per employee not normally distributed

7.20   Work done on a variety of industries in the West Midlands[96] shows that, when related to the number of employees, the frequency with which particular levels of waste arisings occur shows a distribution that is significantly non-normal. The actual distribution of the results may be transformed into a normal distribution by converting one of the variables to a logarithm – this is known as a log-normal distribution.

- The relationship between employee numbers and waste arisings is provisionally described by the equation

    $y = ax^b$

    where

    y is the quantity of waste arising
    x is the number of employees
    a and b are constants.

- In linear form, this equation becomes

    $\log_{10}(\text{waste arisings}) = \log_{10}a + b\log_{10}(\text{number of employees})$

7.21   While this has some advantages for any survey (because, where this holds true, the relationship then obeys the standard statistical rules outlined above), it does complicate the calculation of the minimum sample size still further.

- Data from previous surveys or a pilot survey should be plotted and tested for normality. If the distribution is significantly non-normal, they should be transformed as above and tested again. Statistical advice should be obtained on the next steps, particularly how to calculate sample size and how to deal with confidence limits.

---

[94] Provided the population is normally distributed.

[95] An explanation of *Student's t* distribution and reference tables may be found in almost every statistics text book.

[96] in work carried out for the West Midlands Waste Management Coordinating Authority. See M.E.L. Research: *The West Midlands Industrial Waste Survey*. July 1993.

Coverage and data collection method

7.22 The sample size required, together with the timescale (and the resources available), will dictate the method of data collection – personal visit or postal questionnaire.

> Guidelines for both these methods of surveying are given in chapters 14 and 15.

Deliberate overlaps in sampling

7.23 The WRA in practice should overlap the annual samples in order to check trends.

- Thus the second year's survey should include a sample from the first year's sample to allow comparison: and so on.

Likely errors

7.24 Some estimation errors are inherent in surveys, whatever the data collection method used.

7.25 Other errors are avoidable. The WRA's staff can minimise them if

- they are forewarned about the risks
- they check the raw data returns from time to time with the risks specifically in mind.

7.26 The risks include

- misunderstanding or misinterpreting what 'waste' means
- trying to acquire green credentials by exaggerating the amounts recycled
- inaccurate information about volumes of containers and other such constants
- double counting waste at large waste producers
- omitting waste that is reused or disposed of on site.

Data manipulation

7.27 The inputs and outputs of each class of waste (or identified component of waste) have to be calculated and balanced across each area unit of investigation, and against each facility within the area unit of investigation. The WRA will probably find that an electronic spreadsheet or database handles these calculations better than manual methods can.

> See chapter 16 for more on electronic methods.

Data coding, data entry and data checking

7.28 Choices about data coding, data entry and data checking are to some extent pre-empted by the software the WRA has already chosen.

7.29 In a well-designed survey the major categories – for example, the waste components – will be pre-determined. The database should therefore accept them in coded form.

- If the database design includes numerous free-text fields, the data definitions probably need sorting out.

7.30 Data coding can be done before or after data entry. If coding is done manually (before data entry), the people who do it will need a coding sheet (see paragraph 16.12), but also some knowledge of waste and waste management.

7.31 If coding is done through software, the data are entered direct from the questionnaire. The software

> displays the options

> codes the data

> rejects entries outside a specified range and type, reporting the rejections explicitly

> validates data against pre-determined relationships between fields.

7.32 The data output should be checked for errors and anomalies before it is analysed. Guidance on this is provided in chapter 14.

7.33 Even where software has been used to validate the data, a sample of the data should be checked against the original questionnaires.

- Local knowledge should be brought to bear.

- Seemingly minor shortcomings should not be disregarded: they may – at worst – be the symptoms of major error.

Data analysis

7.34 Effective data analysis will need some statistical expertise. WRAs may therefore consider it more cost effective to carry out the data analysis at regional level. Unnecessary errors are otherwise likely to be introduced into estimates.

7.35 Regional co-operation will be particularly beneficial where small sample sizes would otherwise restrict the analysis. See paragraphs 7.5 to 7.7 above.

The importance of good form design

7.36 Whatever the method of data collection, a form will be needed.

- The form used for personal visits need not be self-explanatory.

- A postal questionnaire should be self-explanatory and easy to complete.

7.37 All forms should be designed with both their purpose and users in mind.

- The design of a form can have a significant effect on:

    a   the number of responses, and

    b   their quality.

- WRAs should consider professional design, or use a desktop publishing package if sufficient expertise is available.

Feedback to the respondents

7.38 During the planning and organisation of the survey, the WRA should, wherever resources permit, provide some feedback to those firms who respond to the survey and especially to those who are visited.

7.39 This feedback is best given in the form of a short written report:

> detailing the findings (type and quantities of waste and processes producing it, major waste components, destinations etc.); and

> summarising any recommendations.

The WRA might also consider whether information from the national waste database for the SIC class or sub-class should be included.

7.40 Such a summary is most easily produced if set up within the standard reporting function of the survey database.

Chapter 8 Future developments

8.1 Chapter 3 (paragraphs 3.56 to 3.60) suggested that the WRA might take a plan period of 8 to 12 years, and divide its discussion of future developments into two:

1. a detailed section dealing with the first 4 to 6 years
2. a more general treatment of the remainder of the plan period.

8.2 In allocating its resources for the plan, the WRA should concentrate on

- increasing the accuracy of the current picture
- assessing the effects of those future changes about which there is a relatively high degree of certainty.

Planned waste management operations

8.3 For information on future waste management operations, the WRA will wish to consult

> the local planning authorities: they will be aware of
>> industry's plans for minerals extraction
>> other potential developments that will affect waste management
> industry, and
>> especially the waste management industry.

Future waste arisings

Household waste

8.4 Population growth and decline affect the amounts of household wastes.

> The WRA can predict the effect of population change on household waste by using data from NHWAP: see paragraphs 12.10 to 12.12 below.

Industrial and commercial wastes

8.5 Changes in industrial and commercial waste arisings are linked to economic activity and changes in employment patterns, as well as to population change.

8.6 In forecasting for industrial and commercial waste, the most useful source will be the survey data on sources of waste. In future years, trends will be derived from the series of annual surveys.

8.7 For a general picture of new industrial growth, the WRA will consult the structure plan. For a more exact assessment of industrial change, the WRA will rely largely upon its close liaison with the development control authorities: the development plan and the waste management plan should go hand in hand.

8.8 The wastes a new firm produces result from its choice of technology: but a reasonable starting assumption is that its wastes are much the same as those of other firms in the same industry.[97]

---

[97] For some industry sectors this assumption will be able to be refined as information becomes available from the government's Environmental technology best practice programme. For further information contact 0800 585794 (telephone) or 01235 463804 (facsimile).

8.9 Information on the national waste database should help the WRA in estimating the amounts of waste from individual companies. No single method, however, will easily or accurately predict future waste arisings: the WRA should combine several sources in its forecast.

- The WRA may find it helpful to build forecasts from
    > estimates of the major individual waste streams, such as sewage sludge and ash from coal fired power stations
    > a combination of other sources of information, such as the industries concerned, HMIP, local authorities' minerals planning departments, and its own local knowledge.

8.10 The forecast of wastes should take account of factors identified during the survey, such as

> projected changes in waste production and management practices in industry

> changes in the area's industrial profile.

8.11 As successive annual surveys are completed, trends in waste arisings, and even waste composition, will become apparent. These direct indications of trends in wastes will increasingly become the WRA's main reference for forecasting waste arisings.

> The forecast should nevertheless still be compatible with other forecasts for the area: e.g. those for population and employment. The WRA should keep in close touch with the local planning authority to ensure this compatibility.

Inbound and outbound wastes

8.12 The WRA will need data for wastes brought into or taken out of its area. It should get these data from

> other WRAs

> data supplied by the waste management industry in the area.

8.13 The WRA should aggregate the data, for each waste type separately, by sending WRA (for inbound wastes) or receiving WRA (for outbound wastes).

8.14 If large quantities of a waste are coming into the area from a single identifiable origin, the sending WRA should be asked to

> assess how much of the waste will be sent into the area, and at what intervals, over the 4 to 6 years of the detailed-plan period

> indicate when the contract (if any) will end or be reviewed.

8.15 Similarly, if waste is being taken out of the area,

> the receiving WRA should be asked how long the receiving operation will continue

> the waste producer (or, for household waste, the waste disposal authority) should be asked when the contract will end or be reviewed.

Recycling and recovery of wastes and waste components

8.16 In forecasting for the 4 to 6 years of the detailed planning period, the WRA should reflect the importance of recycling and recovery by distinguishing the compositions, as well as the quantities, of the major waste types[98].

- The forecast will be more reliable and more useful if the WRA has assessed the strategic factors on which the changes in composition will depend, e.g. the increased recycling and recovery of packaging waste.

8.17 Waste recovery enterprises are subject to much the same economic forces as any other enterprise. Firms engage in materials recovery only if the macroeconomics and microeconomics are favourable.

> Macroeconomic factors include energy costs, raw-material costs, the market for recovered products, and government policies[99] for the encouragement of industrial enterprise.

> Among the microeconomic factors, transport cost is important. A specific waste or waste fraction may be worth recycling only if waste producer, recycler and new user are geographically close.

The WRA should pay full attention to the recovery industry's predictions, but may legitimately contribute its own judgement where the industry's view of macroeconomic factors seems unduly short-term, or (less probably) the industry's assessment of microeconomic factors seems incomplete.

The effect of increased costs

8.18 Government policy – through the landfill tax and strategic pressure for higher waste management standards – will increase the cost of managing waste.[100] This is expected to reduce the quantities of waste and increase the re-use, recycling and recovery of waste materials.

8.19 During the planning process the WRA should

> consider the effects of these policies

> comment on these effects within the plan.

The effects of future legislation

8.20 New legislation can significantly affect the quantities, composition and destination of waste: nor is the effect limited to explicitly environmental legislation. The plan should take account of any new UK or European legislation that affects industry.

- The compliance cost assessment (CCA) that accompanies the consultation document on proposed legislation should provide useful pointers to the legislation's effect, but cannot be taken as conclusive.

Changes in other areas' wastes capacities

8.21 Co-operation and consultation with other WRAs should indicate whether the capacities and capabilities of waste management operations in those WRAs' areas will change; and, if so, the effect on the plan.

---

[98] Waste 'types' means here the 26 waste types listed in chapter 12, paragraph 12.3.

[99] such as Regional Selective Assistance

[100] DoE (1994) *Sustainable development – the UK strategy*, chapter 23, Wastes. HMSO, London.

Chapter 9 Costs and benefits

Resources and burdens

Frequency and intensity of survey

9.1 The WRA should not survey too frequently or too intensively.

9.2 In years 1, 4, 7 (and so on), the WRA should make a full survey of waste arisings. Even then, the WRA should not survey too intensively: it will get a sufficiently accurate picture of current practice without interviewing every producer of waste.

9.3 In the intervening years the WRA should survey a smaller sample of waste producers.

9.4 Every year, however, the WRA will collect from operators of waste management facilities

> a complete dataset about those facilities

> details of waste movements.

Depth..

9.5 The WRA should maintain a clear sense of proportion about the range and detail of the information it asks industry to provide.

9.6 The survey should be clearly targeted. Its purpose is to provide information for waste management planning, not to take a census of the area's industries.

9.7 If the WRA

> seeks data and information from various sources, and

> uses statistically sound sampling,

it should have adequate data for waste management planning without using resources unnecessarily.

.. vs breadth

9.8 But the investigation must not be too narrow, either. It must of course display a sufficiently accurate picture of

> current waste arisings and

> existing treatment and disposal operations.

But it must also look at

> the trends in business and population

so that the WRA can assess any requirement for new or expanded waste management operations. Forecasts should allow for changes in waste composition as well as waste quantity.

9.9 Forecasting the general economic climate ten years ahead is at best an uncertain business: a second forecast balanced on the first will hardly be a model of stability. However accurately the WRA knows the **present** position, it will encounter uncertainties when it tries to predict **future** waste arisings and the manner of their disposal. The attempt must still be made: the predictions should be rolled forward and refined each year. They represent, for the private and public sectors, a key to security and continuity in competitive and environmentally sound waste management.

9.10 Historic cost data for this kind of investigation often do not exist and the WRA may well not be able to make a refined estimate of resource costs at the outset of the investigation.

Reviewing the requirements

9.11 WRAs therefore should review periodically the breadth, depth and frequency of the investigation and the survey in particular in the light of

> the results of previous investigations
> the information needed
> the costs of providing this information
> the benefits accruing to industry through e.g. waste minimisation.

Environmental protection is a key driver

9.12 The WRA, however, cannot carry out its duty to protect the environment unless it

> has information about the management of wastes in its area, and
> is in a position to draw conclusions about their short, medium-term and long-term effects.

Hence, although the WRA must positively manage the investigation's costs, it should **not** do so by arbitrarily omitting essential elements from the investigation.

9.13 To maximise the survey's efficiency, WRAs should work together

> nationally, to plan the survey and decide the levels of accuracy and confidence
> within regions, to devise the sampling strategies.

Comparing data

9.14 Every WRA, building on this co-operation, should check the survey data for consistency with authorities inside and those areas outside its region from which waste is received or to which it is sent.

9.15 Pooling of information increases the chances of finding comparators that are genuinely like-for-like. This is particularly helpful when testing the realism of findings about

> the amounts of waste
> whether wastes are being handled in a way that maximises their availability for re-use
> whether mixing of different waste streams is increasing disposal costs unnecessarily.

9.16 Carefully done, these comparisons will strengthen the WRA's confidence in the completeness and accuracy of its findings, and hence in the advice it offers to industry.

Benefits to industry

9.17 The survey of waste arisings provides an opportunity to

> let industry know more about the quantities and costs of its wastes
> exemplify the return on investment from waste minimisation
> advise industry about waste management options it may have overlooked.

9.18   Wherever possible, the WRA will also be feeding back to firms a summary of the findings from the visit: see chapter 7, paragraphs 7.38 to 7.40 above.

Improving waste management and waste minimisation

9.19   Industry's work in reducing pollution and minimising waste at source is likely to affect waste arisings. At the level of a whole industry, or nationally, the phenomena will be complex and hard to interpret.

1 In the **short** term, requirements by, for example, HMIP for additional abatement may begin to alter the content of waste streams, increase the quantities of waste, and produce new wastes.

2 In contrast, reuse and recycling should start to pull in the opposite direction: that is, to **reduce** the quantities of waste to be disposed of.

3 In the **longer** term, cleaner technologies and more careful materials management should also reduce the national demand for disposal capacity.

9.20   Given these difficulties of prediction at the whole-industry or national level, the WRA should be equipping itself to take into account the individual capability of each company for waste minimisation, reuse, recycling and recovery.

9.21   The WRA, besides

a estimating waste arisings from industry and commerce, and

b getting data about packaging waste and recycling,

should also use the survey to

c provide a baseline against which firms, and the WRA, can measure subsequent waste minimisation

d advise industry on waste management options

e advise industry about the correct waste descriptions to use on duty of care transfer notes

f recommend when the waste producer might profitably separate a mixed waste stream

g stress that waste audit is an essential precursor to waste minimisation

h identify potential for waste minimisation, and

i indicate opportunities for reuse and recycling.

9.22   WRA staff should be capable of handling all of **a** to **i** in face-to-face discussion with industry. If a member of staff is not a specialist, he should be given training.

The importance of weighing waste

9.23   If industry and commerce know the quantities and unit costs of the waste they produce, they are more likely to be motivated towards waste minimisation.

9.24   Within the constraints of practicability, the WRA should encourage all waste producers to

- measure, systematically, the weight of materials in each process waste stream, and thus
- assess the costs of waste[101] more accurately
- pass these data to the WRA.

Identifying the process

9.25 Wherever possible, the WRA should identify a waste stream with a process (rather than merely with a factory). Given this

- the WRA will be able to improve its estimates of waste arisings
- waste producers will be more capable of
  > identifying opportunities for improved waste management
  > attributing waste management costs to individual operations.

---

[101] Beside the cost of waste management, the cost of wastes also includes
> the cost of the materials wasted
> the share of the processing costs (e.g. energy) that these materials represent.

PART II – PRACTICE

Chapter 10 Waste movements study

10.1   This chapter considers the data to be obtained from

> records and returns from waste management facilities, and
>
> details from deposits of waste.

Data from these sources should enable the WRA to derive information about total waste receipts for each type of waste.

- All operations, whether or not they are licensed, should keep records of the quantity, nature and origin of the wastes received.[102]

Licensed facilities

10.2   In addition, licensed facilities must[103] keep **for each type of waste** records of

> its quantity and origin; and, where relevant
>
> its destination, frequency of collection, mode of transport and treatment method.

This information must be made available to the WRA, if requested.

10.3   All waste management licences should contain limits on the quantities of individual categories of waste[104]. Waste management licences should contain the necessary licence conditions so that the origins and quantities of each type of waste are reported to the WRA.

10.4   Each licence holder should report quarterly figures for

> input of wastes by type and origin[105]
>
> output and destinations of
>
>> wastes and
>>
>> materials derived from wastes
>
> stocks of waste.

10.5   Licence holders at landfill sites should also report

> annual figures for the consumption of air space.[106]

---

[102] Paragraphs 14(2) and 14(3) of Schedule 4 to the Waste Management Licensing Regulations 1994, as amended by Regulation 3 of the Waste Management Licensing (Amendment etc.) Regulations 1995 (SI1995 No.288), except activities that are exempt from licensing (except those metal recycling sites that are exempt), from the requirement to keep specific, prescribed records. Article 14 of the Waste Framework Directive however requires all such operations to keep records and to make them available to the competent authority on request.

[103] Paragraph 14 (1), Schedule 4 to the Waste Management Licensing Regulations 1994

[104] Paragraph 6 of Schedule 4 to the Waste Management Licensing Regulations 1994 transposes into UK law requirements of the Framework Directive and specifies matters to be included in a waste management licence as, *inter alia*, 'the types and quantities of waste'.

[105] for further detail on the meaning of 'origin', see paragraph 10.32

[106] required by Waste Management Paper 4 (at appendix A, Checklist – Key points for licensing: part 2, column headed 'Monitoring', item 'Void space')

10.6 The WRA should analyse the numbers for each facility (including landfills) that receives waste. The facility's inputs and outputs of waste should be categorised by sources and destinations: see paragraph 10.12 below.

10.7 To check the accuracy of these figures and those that will be obtained from the survey, the WRA should visit licensed waste operations to witness deposits of various loads. It should

> classify each load[107], and
> record
>> the name and address of the waste producer
>> the name and address of the waste disposal contractor
>> the quantity.

Sites exempt from waste management licensing

10.8 Waste management operations registered as exempt, or those that are subject to other pollution control regimes, also deal with significant quantities of wastes, using methods such as landspreading.

10.9 These operators should[108] keep similar records to licensed facilities and the WRA should:

> for sites that are registered as exempt, request these data are supplied quarterly
> for operations that are subject to other pollution control regimes, request these data at the same frequency from the competent authority, or if they are not available, obtain them from the operator.

The WRA will thus be able to

> provide a complete picture of waste movements
> advise on the suitability of the disposal techniques for the wastes concerned.

Licensed and exempt sites

10.10 At both kinds of site – exempt and licensed – these deposit-based surveys provide valuable additional information about

> the classification of wastes
> the effects of particular types of waste at the particular site (e.g. constituents in the waste related to components in leachate).

This kind of information should be fed back to industry: see chapter 7, paragraphs 7.38 to 7.40.

10.11 At both kinds of site, the WRA should check some of the duty of care transfer notes and consider whether it can obtain additional, useful information from them and, if not, discuss with the licence holder how the information contained on the notes may be

---

[107] Using the national waste classification system (see paragraph 6.54 et seq.).

[108] Under Article 14 of the Waste Framework Directive (see footnote 102).

Waste flows

Input and output totals

10.12 For each waste facility and for every type of waste, inputs and outputs should be calculated[109]

> for every **area** from which waste is received or to which waste is taken.

Internal flows

10.13 Where waste movements are **entirely within** the WRA's area (internal flows), the **local authority area** of origin and destination is sufficient identification.

External flows

10.14 Where the origin or destination is **outside** the WRA's area but within the UK (external flows), the name of the sending WRA or receiving WRA will usually be enough.

International flows

10.15 If waste is imported into the UK (international flows), the country of origin is enough for waste management planning purposes.

Double counting

10.16 Most waste management facilities, but especially waste transfer stations, will have waste outputs as well as inputs.

> Even at landfill sites, waste metal or below-specification capping or lining material may occasionally be removed from site.

10.17 So, to avoid double counting, waste flows **between** waste facilities, including those receiving recovered materials for recycling, e.g. glass cullet, should be separately identified.

- These flows should be shown in the waste inputs and outputs of the site; but they should **not** be included in the waste flows of the area under consideration, **except** when the source or destination is in another area. (For example, if the area that is the subject of the survey is a county, then only outputs from waste facilities outside the county that are inbound and those from waste facilities within the county that are outbound have a net effect on the area.)

Flow charts

10.18 To obtain a complete picture of waste movements within, into and from an area, the authority should build up a series of flow charts. Each chart should show the flows of different wastes (and waste components where useful) between waste management facilities.

10.19 These flow charts should be supported by detailed calculations of the waste flows.

- Because they entail repetitive work, these calculations are best done in a spreadsheet or database package.

- Within the package, the application should be set up to show the inputs and outputs for each type of waste (i.e. each of the 26 headings at paragraph 12.3) at each waste management facility. Changes in storage must be taken into account.

---

[109] The results of these calculations do not need to be published (and may, in any case, be commercially confidential). The method used, however, should be given in a supplementary volume dealing with the technical aspects of the survey (see paragraph 4.18).

Non-controlled wastes

10.20 The WRA should include **non**-controlled wastes because they can affect the disposal of controlled wastes: see paragraphs 5.29 to 5.30 above.

> For non-controlled wastes, fewer details are likely to be needed than for controlled wastes.

> No one can be compelled to supply information about non-controlled wastes.

> For agriculture, non-controlled wastes will include, e.g. animal faecal matter put to beneficial use on the farm.

Using the results – balancing supply and demand

10.21 For **each type** of waste listed at paragraph 12.3, the capacity needed in each year should be calculated as

> the forecast waste arisings in the area[110] **plus** waste brought into the area **minus** waste taken out.

10.22 The required capacity for **that** waste over the plan period is then

> the sum of the capacities needed for that waste in each of the years of the plan period.

10.23 The required capacity for **all wastes** over the plan period is then

> the sum of the capacities needed for each waste in the plan period.

10.24 Recycling may provide an outlet for all or part of a waste stream: this turns on whether the recycled products find established markets. If recycling is succeeding, the projection of required capacity should be **reduced** by the amount of waste the WRA can reasonably assume will go to recycling.

10.25 Some wastes may be destined for incineration, or other treatment. If so, the WRA should allow within the plan for disposal of the treatment residues.

- Residues from waste incineration are only about 10% by volume of the original waste. Even so, they may still represent a substantial waste arising and will need a considerable volume of secure landfill.

Improving records of waste received at waste management facilities

10.26 None of this is feasible, however, unless records of waste movements at waste management facilities are **all** of a sufficiently high standard.

10.27 Waste Management Paper 4[111] gives statutory guidance on

> the need for waste inputs to be weighed

> the type of return which licence holders should be required to make.

---

[110] This figure will be produced from the survey of industrial and commercial waste producers and forward estimates based on likely trends in waste arisings.

[111] DoE. Waste Management Paper No 4, *Licensing of Waste Management Facilities*. London: HMSO: 3rd edition 1994. ISBN 0 11 752727 0.

Weighing should be the normal method

10.28   Waste inputs should normally be measured by weight. The licence holder should provide and use weighing facilities[112]: this usually means some form of weighbridge.

- The WRA should ensure that every waste management licence includes conditions to this effect, **except** where they would represent an **unreasonable** burden on the licence holder. Reasonableness depends on the facts of the case, but should be tested against

    > the pollution *potential* of the site, measured by the quantities and types of waste it receives

    > the site's significance in the area's waste disposal capacity as a whole.

Exceptions

10.29   Thus, in general, the only exceptions will be

> sites licensed to take excavated inert materials, such as subsoil from a single quantified source

> small sites (where small is interpreted as not more than 5,000 tonnes per year) that are licensed to take inert wastes only. At these sites, the licence should require the licence holder to use an external weighbridge.

Using weight and other data to check waste arisings

10.30   Weighing and proper recording of waste inputs will substantially improve the accuracy and detail of the data from the waste management facilities.

10.31   The WRA should use

> these records, along with

> data from its inspections of consignments of waste at the facility

to correlate the facility's waste input data with parallel data derived from the WRA's survey of waste producers. These parallel data include

> the waste producer's industrial class under the SIC

> the amount and composition of the wastes his operations produce.

10.32   In addition, each licensed waste management facility must[113] record the origin, quantity and type of wastes it receives.

> The **origin**[114] should be shown in a way that identifies the company that produced the waste and the company's location.

> The **type** of waste should be taken from the national waste classification scheme: see paragraphs 6.54 to 6.59.

---

[112] paragraph 4.18 of WMP4

[113] Under the Waste Management Licensing Regulations 1994, Schedule 4, paragraph 14.

[114] This will usually mean the name and full address (including the post code) of each waste producer. In some circumstances the name and postcode only may be acceptable.

Timing of the waste movements study

10.33 The information gathered from waste facilities on the types, quantities and sources of waste they receive and transfer out (the *waste movements study*) will

- provide a valuable check on the results from the survey of waste arisings
- inform that survey's staff about
    > the size of the populations to be sampled
    > the location of some of the members of the populations.

10.34 Therefore

- the waste movements study is probably most useful if carried out at the **outset** of the annual survey of waste arisings.

10.35 The results will also allow the WRA to examine the feasibility of upgrading site returns to provide, routinely, the information the WRA will need.

Waste movements study v. producer survey: efficiency and effectiveness

10.36 Waste producers, the waste management industry and waste regulators will need to exert sustained efforts to ensure that adequate data on weights, types and origins of waste are provided to, and recorded at, waste management facilities.

> Collecting information in this way will nevertheless involve substantially less effort than obtaining similar waste arisings data from visits to waste producers.

10.37 The survey of waste producers, however, still offers benefits: in particular

> it is a more effective way of obtaining accurate data on waste components and the processes producing waste

> it provides an opportunity to advise on waste minimisation recycling re-use and recovery.

The two approaches are therefore complementary and both should be used.

Chapter 11 Census of waste management facilities

Introduction 11.1 This chapter specifies the data the WRA should seek to collect when it takes its census of existing treatment and disposal facilities, their nature and capacity.

Data sources 11.2 The WRA will already have many details of existing licensed facilities and some for registered sites.

Sites under other regimes 11.3 Where some other control regime[115] is substituted for waste management licensing, the WRA should get the data from the authority (e.g. HMIP) or, if the data are not readily available by that route, from the public register or, if not available there, from a site visit.

- Even operations registered as exempt from licensing may still be dealing with substantial quantities of wastes, particularly when considered together: if necessary, the WRA should visit them to obtain the data.

Logistics of site visits 11.4 A site visit improves the accuracy of the information: the visit may be combined with, for example, a site inspection. Some of the information may already be available from the study of waste movements, together with the site returns.

Data according to type of site 11.5 Table 11.1 sets out the data the WRA will need. The table assumes four types of site, with a slightly different dataset for each:

    a treatment plants: this includes incineration and anaerobic digestion, but **excludes** composting facilities and recycling centres, which are in **c** below

    b landfill sites

    c composting facilities and recycling centres (including MRFs, vehicle dismantlers, scrapyards and others)

    d transfer stations.

Most of the data items, however, are common to all four types.

---

[115] such as IPC under part I of EPA90

Table 11.1

Data items to be collected in the census

Key
✓ = this item to be collected at this kind of site
• = this item does **not** apply at this kind of site

serial	data item [units]	data type	a Treatment plant	b Landfill	c Composting facil. or recycling centre	d Transfer station
1	Location [OS co-ordinates of main entrance]	numeric	✓	✓	✓	✓
2	Type of site or process	text [as pick-list]	✓	✓	✓	✓
3	Date opened	date	✓	✓	✓	✓
4	Remaining useful life [years]	numeric	✓	✓	✓	✓
5A	Probable closure date	date	✓	✓	✓	✓
5B	Probable completion date	date	•	✓	•	•
6A	Classes of waste actually being dealt with	text [as pick-list]	✓	✓	✓	✓
6B	Quantity of waste in each such class [tonnes/year]	numeric	✓	✓	✓	✓
7A	Classes of waste the site could deal with	text [as pick-list]	✓	✓	✓	✓
7B	Quantity of waste in each such class [tonnes/year]	numeric	✓	✓	✓	✓
8A	Space capacity by class, 7B - 6B [tonnes/year]	numeric	✓	✓	✓	✓
8B	Total remaining capacity [tonnes]	numeric	•	✓	•	•
9A	Recycled output[116]: type	text [as pick-list]	•	•	✓	✓
9B	quantity [tonnes/year]	numeric	•	•	✓	✓
9C	destination [OS co-ordinates]	numeric	•	•	✓	✓
10A	Final waste: type	text [as pick-list]	✓	•	✓	✓
10B	quantity [tonnes/year]	numeric	✓	•	✓	✓
10C	destination [OS co-ordinates]	numeric	✓	•	✓	✓
11	Pollution control measures	text [as pick-list]	✓	✓	✓	✓
12A	Energy recovery - type	text [as pick-list]	✓	✓	•	•
12B	Energy recovery – amount [MWhre./annum]	numeric	✓	✓	•	•
12C	Energy consumed – amount [MWhre./annum]		✓	✓	✓	✓
13	Policies on waste acceptance	text [as pick-list]	✓	✓	✓	✓
14	Policies on availability to waste producers, hauliers and disposal contractors	text [as pick-list]	✓	✓	✓	✓

[116] It is assumed that, where wastes are recycled at a landfill, the operation will be covered by a separate licence: if it is not then data on recycled wastes should be collected for landfills.

Chapter 12 Key waste arisings and their estimation

12.1 This chapter

> discusses how to get the data of waste arisings in a useful and common format

> lists the types of waste that should be separately reported

> outlines how to estimate waste arisings for each of the major types.

Types of waste to be reported separately

12.2 The investigator should collate data on wastes at the Division level of the Standard Industrial Classification 1992: see table 6.2.

12.3 For reporting, the investigator should group the data under the following 26 headings.

A1 Collected household waste[117]

A2 household waste delivered to civic amenity sites[117]

A3 household waste delivered to other bring systems (e.g. bottle banks)[117]

A4 household waste dealt with at home[117]

A5 litter, street sweepings and other household waste

B   commercial wastes, excluding H and I

C   industrial wastes, excluding D to G

D1 construction and demolition wastes, excluding D2 to D4

D2 excavated soils and sub-soils, (excluding D3)

D3 contaminated soils

D4 asphalt road planings

E   special waste

F   power station ash and flue gas desulphurisation sludges

G   incineration residues (excluding special wastes)

H   healthcare wastes, including clinical wastes

I   sewage sludges

J   sewage not disposed of to the sewerage system

K   tyres[118]

---

[117] Household waste **ex**cludes
> medical and veterinary wastes – see section H
> sewage not disposed of to the sewerage system – see section J.

[118] The survey of industrial and commercial premises should wherever possible identify separately the quantity of tyres and the methods used for their disposal.

    L   scrap cars, trailers, railway rolling stock

    M   packaging[119]

    N   fragmentiser residues

    O   food processing

    P   water treatment plant sludges

    Q   radio-active wastes

    R   mine and quarry wastes[120]

    S   agricultural wastes.[120]

The rest of this chapter discusses some of the most significant of these headings individually.

A1-A5 Household wastes

12.4 Household wastes are defined in section 75(5) of EPA90, and prescribed in the Controlled Waste Regulations 1992[121]. The definitions are set out in box 12.1. They include domestic and civic amenity wastes, litter and street sweepings.

12.5 In estimating the composition and arisings of household waste, the WRA staff will probably have to incorporate information from several sources: they must therefore examine each source critically, so as to establish its reliability and accuracy.

12.6 Useful information may be available from

> district councils (where it will derive from work done for their recycling plans and as waste collection authority)

> the waste disposal authority

> contractors with responsibilities for collection or subsequent management and disposal of household waste in the area

> estimates from the study of waste facilities

> surveys conducted for other purposes – e.g. estimating the likely inputs to a materials recycling facility (MRF); and

> the database set up under the National Household Waste Analysis Programme (NHWAP): see paragraph 12.10 below.

12.7 Waste from the recycling centres at civic amenity sites should be included as household waste. Where the quantity of these wastes is not known, the WRA should arrange for them to be weighed.

12.8 The WRA staff should compare the quantity of civic amenity waste with the quantity of collected household waste. Civic amenity waste is usually

---

[119]The survey of industrial and commercial premises should wherever possible identify separately the amounts of packaging wastes and the methods used for their disposal.

[120]Only those mine and quarry wastes and agricultural wastes that are controlled need be fully estimated and reported. Non-controlled wastes should be taken into account where they affect the availability of options for controlled wastes and the extent that this may occur should be reported.

[121]SI 1992 No. 588

> about one third of the collected household waste figure if the collection system does **not** use wheeled bins

> about one quarter of the collected household waste figure if wheeled bins **are** used.

12.9 Where different fractions of the household waste are currently, or might in future be, separately collected, WRA staff should record the type and quantity of each fraction separately.

Box 12.1   *Wastes defined and prescribed as household waste*

Household waste is defined in section 75(5) of the Act and, subject to any Regulations made under section 75(8),

'means waste from-

- domestic property, that is to say, a building or self-contained part of a building which is used wholly for the purposes of living accommodation;
- a caravan (as defined in section 29(1) of the Caravan Sites and Control of Development Act 1960) which usually and for the time being is situated on a caravan site (within the meaning of that Act);
- a residential home;
- premises forming part of a university or school or other educational establishment;
- premises forming part of a hospital or nursing home.'

and is also prescribed (Controlled Waste Regulations 1992) as waste from

- a hereditament or premises exempted from local non-domestic rating by virtue of:-
    - in England and Wales, paragraph 11 of Schedule 5 to the Local Government Finance Act 1988 (places of religious worship etc.);
    - in Scotland, section 22 of the Valuation and Rating (Scotland) Act 1956 (churches etc.)
- premises occupied by a charity and wholly or mainly used for charitable purposes
- any land belonging to or used in connection with domestic property, a caravan or a residential home
- a private garage which either has a floor area of 25 square metres or less or is used wholly or mainly for the accommodation of a private motor vehicle
- private storage premises used wholly or mainly for the storage of articles of domestic use
- a moored vessel used wholly for the purposes of living accommodation
- a camp site
- a prison or other penal institution
- a hall or other premises used wholly or mainly for public meetings
- a royal palace

and is waste

- arising from the discharge by a local authority of its duty under section 89(2) of EPA90. This prescription covers all waste collected by a local authority in the course of discharging its duty under section 89(2)(a). It may, for example, include litter and dog faeces.

National Household Waste Analysis Programme

12.10 To help waste regulation authorities in estimating the quantity and composition of household waste, the Department of the Environment[122] funds the development of a National Household Waste Analysis Programme (NHWAP).

> The first results of phase 2 of the NHWAP project were published in 1994[123]

---

[122] with contributions from the Department of Trade and Industry and the Industry Council for Packaging and the Environment (INCPEN)

[123] CWM 082/94 *National Household Waste Analysis Project Phase 2. Report on composition and weight data.* AEA Technology and Aspinwall & Co for Department of the Environment 1994: available from WMIB (see footnote 124)

> The Department intends to continue the NHWAP, which should provide progressively more precise information on household waste arisings and composition.

12.11 A database at the Waste Management Information Bureau[124] holds the results of the composition analyses so far carried out on household waste under the NHWAP.

- For waste management planning purposes, WRAs and local authorities may access the database free of charge.

12.12 The database is being expanded to incorporate information on industrial and commercial waste: see paragraphs 6.42 to 6.46.

Including all wastes from the household

12.13 The quantity and composition of the household waste produced in an area is broadly dependent on the size and socio-economic class of the population.

12.14 The quantity and composition of **collected** household waste is additionally influenced by the methods chosen for managing household waste in the area. The factors include

> the publicity given to recycling
> the convenience of bring facilities
> above all, the size and type of container provided for the collection of household waste: for example, the use of wheeled bins increases the weight of collected household waste substantially – sometimes by one third or more.

12.15 Estimates of the quantity and composition of household wastes should therefore include all wastes arising from households in the area, whether the waste is

> collected
> taken to civic amenity sites
> taken to bottle banks and paper banks
> taken to scrap metal recycling facilities
> disposed of on the premises – eg. on an open fire, in a stove or by composting.

Questionnaire survey by the WRA

12.16 The WRA should estimate the amount and composition of the wastes being dealt with by each of these separate routes. To do so, the WRA may wish to send questionnaires to a sample of households.

- The WRA should also consider whether it should concurrently sort and weigh the wastes from the sample. This method of checking weight and composition may be particularly useful when no original data exist.

12.17 Guidance on methods for this kind of survey is given in several of the research reports published under the Controlled Waste Management (CWM) Research Programme.[125]

---

[124] Waste Management Information Bureau: 'phone 01235 463162

[125] A list of CWM reports and the reports themselves are available from the Waste Management Information Bureau at NETCEN. Telephone 01235 463162 or facsimile 01235 463004.

Other household wastes

12.18 The WRA will need to include household waste **not** produced from domestic premises (see box 12.1).

12.19 Premises such as residential homes, hospitals (see paragraphs 12.52 and 12.56), schools and universities may need to be visited individually.

- As with commercial and industrial waste, the WRA staff should give advance warning of
   > the rationale for the investigation
   > the data items the WRA needs.

B   Commercial waste

12.20 Commercial waste is defined and prescribed as including waste from premises of the types set out in box 12.2. Chapter 13 gives details of the SIC(92) categories relating to commercial wastes. These arisings are often more difficult to estimate than household waste. Their sources are more varied; so also are the methods for managing them.

12.21 Commercial waste may be

> collected by the local authority contractor, who may charge for the service

> disposed of at a dual-purpose site (a civic amenity site combined with a transfer station)

> collected by a private waste disposal contractor.

Box 12.2   *Wastes defined and prescribed as commercial waste or not commercial waste*

Commercial waste is defined in section 75(7) of the Environmental Protection Act 1990. Subject to any Regulations made under section 75(8) of the Act, it means waste from premises used wholly or mainly for the purposes of a trade or business or the purposes of sport, recreation or entertainment excluding household waste; industrial waste; waste from any mine or quarry and waste from premises used for agriculture, within the meaning of the Agriculture Act 1947 (in Scotland, the Agriculture (Scotland) Act 1948); and waste of any other description prescribed by regulations made by the Secretary of State for the purposes of this paragraph [ie paragraph (d) of section 75(7)].

Waste to be treated as commercial waste by virtue of regulations made under section 75(8)

Waste from an office or showroom

Waste from a hotel

Waste from any part of a composite hereditament which is used for the purposes of a trade or business

Waste from a private garage which either has a floor area exceeding 25 square metres or is not used wholly or mainly for the accommodation of a private motor vehicle

Waste from premises occupied by a club, society etc.

Waste from premises (not being premises from which waste is by virtue of the 1990 Act or of any other provision of the Regulations to be treated as household waste or industrial waste) occupied by a court; a government department; a local authority; a body corporate or an individual appointed by or under any enactment to discharge any public functions; or a body incorporated by a Royal Charter

Waste from a tent pitched on land other than a camp site. 'Camp site' is defined in regulation 1(2).

Waste from a market or fair

Waste collected under section 22(3) of the Control of Pollution Act 1974. Section 22(3) of the 1974 Act provides that a local authority may enter into an agreement for the cleaning of any land in the open air to which members of the public have access. This prescription covers all waste collected by a local authority in the course of cleaning land under section 22(3). It may, for example, include dog faeces.

Waste NOT to be treated as commercial waste

Sewage sludge or septic tank sludge which is treated, kept or disposed of (other than by means of mobile plant) within the curtilage of a sewage treatment works as an integral part of the operation of those works

Sludge which is supplied or used in accordance with the 1989 Regulations

Septic tank sludge which is used in accordance with the 1989 Regulations.

Survey method

12.22 The survey should follow the guidelines given in chapters 13 to 16. It should aim to quantify the amounts of different types of waste that are

> recycled
> reused
> recovered
> sent for disposal.

12.23 The WRA staff should

> arrange the weighing of waste from individual premises
> cross-check the information obtained from this survey with the quantities of wastes received at transfer stations and disposal sites
> sub-divide packaging wastes into their main types.[126]

C. Industrial waste (excluding D: construction and demolition waste)

12.24 The definition of industrial waste is set out in box 12.3. Additional wastes prescribed as industrial wastes[127] are listed in Appendix B. Chapter 13 gives details of the relevant SIC(92) codes for industrial waste.

Box 12.3 *Definition of industrial waste*

Industrial waste is defined in section 75(6) of the Act. Subject to any Regulations made under section 75(8), it means 'waste from any of the following premises-

(a) any factory (within the meaning of the Factories Act 1961);

(b) any premises used for the purposes of, or in connection with, the provision to the public of transport services by land, water or air;

(c) any premises used for the purposes of, or in connection with, the supply to the public of gas, water or electricity or the provision of sewerage services; or

(d) any premises used for the purposes of, or in connection with, the provision to the public of postal or telecommunications services.'

12.25 Industrial wastes (as defined) are produced from a very wide range of activities, and are immensely varied: estimation is inherently more difficult than for commercial wastes.

• The WRA must nevertheless carry out a properly constructed survey.

> Industry, existing and new, might otherwise have cause to doubt whether its planned waste production would be adequately dealt with.

Guidelines for such a survey are given in Chapters 13 to 16.

---

[126] Packaging wastes should be sub-divided as required by the Directive: that is by material type (plastic, paper, card, glass etc.) and by usage (primary packaging, secondary packaging etc.)

[127] Section 75(8) of the Environmental Protection Act 1990: Regulation 5(1) and Schedule 3 to the Controlled Waste Regulations 1992

12.26   Industrial wastes as a general category are heading C (see paragraph 12.3). The WRA staff should ensure that the related waste types – construction and demolition waste, for instance – are dealt with under their own separate headings D1 to D4.

Separate consideration of large waste streams

12.27   Some waste streams are produced in large quantities at a single location. They include

> power station ash
> blast furnace slag
> other steel-making wastes
> water treatment plant sludges.

A single change in the process or the legislation could substantially alter the demand for future waste disposal capacity. Hence these streams should be separately considered in the survey.

Separate consideration of other waste streams

12.28   Other specific waste streams that should be identified and quantified separately include

> packaging from industry
> incinerator residues
> fragmentiser wastes
> tyres
> dredgings.

D1 Construction and demolition wastes (excluding excavated materials)

12.29   These wastes arise mainly from the private sector, although some will be produced by local authorities.

12.30   At least some of the waste from every construction project will be deposited at a waste site. Site records will probably be the most complete guide to the existence of construction projects.

Data Sources

12.31   When the disposal site records have identified the construction projects, the WRA staff may be able to get quantitative data from

> the construction firms
> demolition contractors
> haulage contractors
> local authorities' environmental health departments (on, for example, the outputs from concrete crushers and other plant used to treat waste on site).

These approaches should also bring the WRA's notice other waste arisings that are being recovered or recycled.

Recycling construction and demolition wastes

12.32   Waste arisings in this category are substantial. During the investigation, the WRA should seek to make the industry more aware that some of these wastes – especially wastes from demolition – can be profitably reused and recycled. Details of recycling and potential uses for recycled materials are given in a report produced for DoE[128].

---

[128] Howard Humphries and Partners, (1994), *Managing Demolition and Construction Wastes*. HMSO, London. ISBN 0 11 7529729

12.33 More particularly, the WRA, in its discussions with local authorities and private developers should encourage them to specify suitable secondary materials for their civil engineering and building works.

12.34 The terms of a demolition contract should encourage separation.

> Slates, tiles, and some bricks can be easily reused.

> Glass and metal are readily recyclable.

> Wood and plastic may be recyclable, or can be burned to produce energy.

> Road planings, concrete, and bricks have potential for use as secondary aggregates.

12.35 The WRA staff should estimate separate quantities for

> tiles and bricks for reuse

> road planings

> secondary aggregates such as crushed concrete and crushed bricks.

12.36 Both these (construction and demolition) wastes and excavated materials (described in the following section) should be dealt with in a section of the plan separately from other wastes (see chapter 4).

D2 Excavated materials (engineering spoil)

12.37 Construction projects often produce a large quantity of excavated materials. These are referred to as engineering spoil. Where such spoil is neither disposed of through *cut and fill*[129] nor recycled it may consume substantial landfill void space.

12.38 Close to water sources, or in areas where water resources are highly vulnerable, however, these excavated materials may be the only waste acceptable for the restoration of mineral workings.

12.39 Such wastes are practically speaking inert: they should present few disposal problems.

12.40 Nevertheless, their disposal in large quantities at landfill sites that also take biodegradable waste might be seen as affecting the sustainability of landfilling and the assessment of completion. Hence the WRA must assess the arisings so that appropriate voids can be planned and provided.

12.41 As with construction and demolition wastes, information on the arisings will come mainly from records at waste management facilities, but partly from

> construction companies

> demolition contractors

> haulage contractors.

---

[129] Material excavated from one part of a construction site and deposited at another part of the same site, particularly used in road building.

12.42   Cut and fill operations should be ignored where there is no net impact on other controlled wastes. If, however, excavated material is deposited outside the construction site boundary, it becomes the subject of a waste management licence (or is registered as exempt).

- Such licensed or registered waste should be included in the survey.
- But since by definition its net effect on capacity is nil, the quantities should be separately identified.

12.43   Neither contaminated soils nor river and canal dredgings[130] should be included in this category.

D3 Contaminated soils

12.44   Contaminated soils arise as **wastes** primarily from the redevelopment of derelict land or old industrial sites.

12.45   Such wastes may arise in substantial quantities and with varying pollution potential, particularly if they have not been treated on site to reduce their volume and/or their toxicity. They may only be acceptable at a limited range of waste management facilities.

12.46   Details of recent and current arisings, and the sites at which they were deposited (where these are within the WRA's area), should be available to the WRA from its own or site records. Co-operation with other WRAs should establish recent and current arisings that have been deposited outside the WRA. For information on future arisings, the WRA should approach

> the construction industry

> the local planning authorities.

G   Incineration residues

12.47   Incineration residues arise principally from the incineration of municipal wastes. A much smaller contribution comes from sewage sludges and clinical wastes, and a very minor quantity from hazardous wastes.

12.48   Incineration residues should have a very low carbon content: they thus produce very small amounts of methane.

12.49   Operators usually recover ferrous metal from MSW incinerator ash. They should always have considered whether to do so, even though recovery is unlikely to reduce waste volumes significantly.

> Bottom ash from MSW incineration is generally about 10% of the original waste volume, and about 25 to 30% of its weight.

> Wastes from pollution control including fly ash represent a further 3 to 4% of the waste input and must also be catered for.

12.50   Incineration residues may have significant pollution potential due to leaching

> toxic heavy metals

---

[130] The Controlled Waste Regulations 1992 prescribe waste from dredging operations as industrial waste.

> soluble compounds such as chlorides.

12.51 Landfill leachate from incineration residues can thus be difficult to treat to a standard fit for discharge. The WRA will need to consider what disposal sites and methods will be adequate.

H Health Care wastes

12.52 This category includes wastes from

> hospitals

> doctors' and dentists' surgeries

> health centres

> nursing homes

> veterinary surgeries.

It also includes

> community generated clinical waste, where this is collected separately.

12.53 In law these may be variously defined as household, industrial or commercial waste.

> Some health care wastes (principally those containing prescription-only medicines, or substances with a flash point at or below 21°C) will also be special waste.

12.54 Options for recycling are limited by the nature of the waste and its potential hazards.

12.55 The classification and disposal options for this type of waste are given in Waste Management Paper 25.[131]

12.56 Clinical waste should be separately recorded from other general wastes from hospitals. It should be categorised according to the disposal options that might be acceptable. Hospital procedures should be examined at first hand to ensure that

- separation practice aligns with the disposal options available for different types of waste

- the amount of waste categorised as hazardous is minimised by careful separation.

I Sewage sludges

12.57 Records of sewage sludge arisings and disposal routes may be obtained from the statutory sewerage undertakers.

12.58 Sewage sludge may be

> spread on land[132]

---

[131] The Department intends to publish a revised version of WMP 25 by Spring 1996.

[132] When spread on agricultural land, the operations are subject to the provisions of the Sludge Use in Agriculture Regulations 1989, SI 1989 No. 1263 (as amended). Sewage treatment works must keep records, including: quantities, composition and properties of the sludge and details of where it was spread.

> incinerated

> landfilled

> used in composting.

12.59   Some sewage sludges are sent for sea disposal: this ceases at the end of 1998. The WRA should thoroughly assess their sewage undertakings' proposals for alternative disposal methods.

- The WRA can estimate the mass of sludge, **as dry solids**, by assuming a rate of 0.26 tonnes dry solids/person/year for the population served by the treatment works.

- The actual weight of waste will depend on the moisture content of the sludge.

- Moisture content varies with the process used. It may typically be anything from 60% for digested, dewatered sludge through to 98.5% for wet sludge from an activated sludge process.

- The WRA staff should estimate both the total weight of arisings and the dry solids content.

Non-controlled wastes

12.60   Most mine and quarry wastes and agricultural wastes are not controlled wastes. Thus planning for dealing with such arisings is outside the scope of waste management planning and this guidance. However, paragraphs 5.29 and 5.30 show how the disposal or recovery of **non-controlled** wastes may significantly affect waste management options for **controlled** wastes. Therefore, non-controlled wastes should be included in the investigation to the extent that they are relevant for waste management planning.

R Mine and quarry wastes

12.61   The Waste Framework Directive[133] does **not** apply to waste from

> prospecting

> the extraction, treatment and storage of minerals

> the working of quarries.

Such wastes are **not** controlled wastes.

- An exception to this is wastes arising from prospecting and exploration for, or production of, onshore oil and gas. These are usually liquids or sludges and are dealt with as controlled wastes.

12.62   However, controlled wastes and mine and quarry wastes may be disposed of in the same site.

- As well as reducing the apparent void available for controlled waste, this may affect the type of waste a landfill is prepared to accept (see paragraph 5.30).

The investigation should therefore also assess current and possible future sites at surface mineral workings where significant amounts of wastes from

---

[133] 75/442/EEC as amended by 91/156/EEC

mines and quarries are, or might be, landfilled at the same site as controlled wastes.

12.63   The extraction of a few minerals, principally china clay, deep-mined coal and slate, generates large amounts of waste. The disposal of such waste is usually an integral feature of the extraction permission; it is separate from any disposal of controlled waste. The minerals planning authorities should be able to provide sufficient detail on locations, waste arisings and the industry's intentions for waste management planning.

12.64   For other mineral extraction, WRA interests will be related to where controlled wastes are, or may be, landfilled in the void. For these, the WRA staff should first get information from the minerals planning authority, who should have data available on

> the number and areas of surface mineral workings
> the number and areas of mineral waste disposal
> their locations
> the areas of the individual permissions or sites
> the name of the operators
> the type of mineral extracted (see box 12.4 for mineral names)
> areas proposed or permitted for new surface workings
> any planning limitations affecting the current or likely future availability of mineral void space for landfill.

and for these sites may also have

> information on the nature and the quantities of minerals waste arising and disposed of with controlled wastes.

WRA staff should also consult their own records for details in relation to licensed sites.

12.65   Where the required information is not available from these sources, the WRA staff should contact the operators direct. Information obtained in this way should be provided to the mineral planning authority.

Box 12.4 *Minerals types used by minerals planning authorities to summarise types of development*

Chalk
China clay
Clay/shale
Coal (open cast)
Coal (deep mine)
Gypsum/anhydrite
Igneous rock
Ironstone
Limestone/dolomite
Sand and gravel
Sand
Sandstone
Slate
Vein minerals
Other minerals

S Agricultural wastes

12.66　The Waste Framework Directive excludes from control only *faecal matter and other natural, non-dangerous substances used in farming*.[134]

- The WRA should estimate even these excluded wastes: land taken for spreading agricultural wastes is unavailable for spreading controlled wastes.

12.67　The local offices of the Ministry of Agriculture, Fisheries and Food can provide details of

> the area of land used for each type of agriculture

> the kinds and numbers of farm animals in the area.

12.68　The WRA should **not** classify a waste as agricultural **unless** it is actually produced at agricultural premises; but **not** all wastes produced at agricultural premises are agricultural waste – see the next paragraph.

Excluding food processing wastes

12.69　Wastes may be produced from food processing at agricultural premises. Where wastes are produced from a food processing factory[135], the food processing wastes from the factory are statutorily[136] industrial wastes. WRA staff should record such wastes in the course of their visits to agricultural premises; but they should classify the data, and output the information, under survey heading O (Food processing waste), **not** under survey heading S (see paragraph 12.3).

Quantifying area needed for landspreading

12.70　Most agricultural waste in the UK is disposed of by spreading it on land.

> Table 12.1, below, gives typical land areas for some animal wastes.

> Areas for other animal wastes can be derived from data in MAFF Reference Book RB 209, combined with the numbers from the WRA's survey.

> Appendix IV of the *Code of Good Agricultural Practice for the Protection of Water*[137] gives guideline figures for the production of excreta by livestock: WRA staff can use the figures to check the survey data.

12.71　The precise area of land per unit quantity of manure or slurry depends on a combination of factors, including

> the nitrogen content of the waste

> the type of land cover

> the slope of the land

> the land's proximity to surface waters.

Different kinds of farm

12.72　Farms should be categorised according to the type of farming activity. The WRA should survey a random sample of each type, using

---

[134] Article 2.1 (iii)

[135] within the meaning of the Factories Act 1961

[136] even though the European Waste Catalogue classifies food processing wastes as agricultural wastes

[137] MAFF, (1991), *Code of Good Agricultural Practice for the Protection of Water*. MAFF Publications, London.

> a questionnaire, with

> a follow-up visit to cross-check.

12.73 The survey should show

> the amount of wastes produced by different farming practices

> the amount of land required to put such wastes to beneficial use.

Table 12.1

Typical land area required for spreading wastes from different livestock[137]

Number and type of livestock	Land area needed for spreading wastes
1 dairy cow (6 month housed)	0.16 ha
1 beef bullock (6 month housed)	0.10 ha
1 pig place (20-90kg)	0.04 ha
1 sow and litters (to 4 weeks)	0.07 ha
1000 laying hen places	2.3 ha
1000 broiler places	1.4 ha

Uses for agricultural wastes

12.74 As well as those agricultural wastes that are recovered by beneficial spreading on land, some agricultural wastes (e.g. poultry litter, straw) are a source of energy when burned, either alone or in combination with non-agricultural wastes.

12.75 Other agricultural wastes (e.g. straw, animal slurry) may serve as feedstock for composting or anaerobic digestion.

---

[137]MAFF, (1991), *Code of Good Agricultural Practice for the Protection of Water.* MAFF Publications, London.

Chapter 13 Selecting industrial and commercial firms for the survey

13.1 This chapter gives practical guidance on how to select samples from the business population.

Profiling the area

13.2 WRAs should examine the profile of industry and commerce in both the region and each WRA area to provide an overview of the population of firms as a whole. Information from the Census of Employment should be arranged in the format set out in table 13.1.

Table 13.1

Area industry profile of information from the Census of Employment

SIC(92) Class		Number of units (premises) and number of employees within each employee size band							TOTAL
		Employee size bands							
		1-10	11-24	25-99	100-199	200-499	500-999	1000+	
15.1	Units	n	n	n	n	n	n	n	n
	Employees	n	n	n	n	n	n	n	n
15.2	Units	n	n	n	n	n	n	n	n
	Employees	n	n	n	n	n	n	n	n
15.3	Units	n	n	n	n	n	n	n	n
	Employees	n	n	n	n	n	n	n	n
15.4	Units	n	n	n	n	n	n	n	n
	Employees	n	n	n	n	n	n	n	n
15.5	Units	n	n	n	n	n	n	n	n
	Employees	n	n	n	n	n	n	n	n
etc...									
TOTAL	Units	n	n	n	n	n	n	n	n
	Employees	n	n	n	n	n	n	n	n

Sub-dividing the population

13.3 The Census of Employment or Business Database listing should be transferred to a database file. This database file can then be used as a source for some of the lists of businesses the WRA will need. In the descriptions that follow, this database file is called the *allfirms file*.

13.4 To achieve greater accuracy in estimation with the effective use of resources, the WRA staff should divide the firms on the list into five groups[138]

Group A   Special waste producers

Group B   20% of industrial firms representing the large industrial employers

Group C   20% of commercial firms representing the large commercial employers

---

[138] This chapter describes the sampling process as if all producers of special waste are to be visited. If the WRA wishes to keep the survey separate from its visits to special waste producers, it should ignore references to the special waste group, Group A.

Group D The remaining industrial firms

Group E The remaining commercial firms.

Reference numbers for waste producers

13.5 Each firm should be allocated a unique *producer reference number*.

- Duplication of the producer reference numbers must be prevented.

  1 A satisfactory way of doing this is to use the unique record number or index reference from the firm's record **in the allfirms file**. The record number can be preserved from the effects of subsequent file creation: make a PRODREFNO field, then copy the allfirms record numbers into it.

  2 If the database designer creates a separate file for each of groups A to E, the record numbers in those files may **not** be unique across the whole set of firms.

Separating industrial, commercial and other waste producers within SIC(92)

13.6 Industrial and commercial waste are statutorily defined and prescribed.[139] Some kinds of business produce industrial waste, others commercial waste.

13.7 The SIC categories of industries that produce industrial waste are listed – at Division level – in table B1 of Appendix B. Industries and trades that produce commercial waste are denoted by the SIC categories listed – again at Division level – in Appendix B, table B2.

13.8 For publication in the plan, industrial and commercial wastes may be aggregated.

Not included in the survey of industrial and commercial waste producers

13.9 Information on some wastes (e.g. contaminated soils) is best **not** produced from the survey. The WRA will instead use data from a combination of different sources – such as the waste movements study, local knowledge, and contact with other agencies.

13.10 The next three paragraphs discuss wastes that should be treated in the survey as separate from industrial and commercial waste. Their SIC codes are also listed in Appendix B.

Construction industry

13.11 Construction industry wastes should be separated from the industrial and commercial survey; the quantities – which are very large – should be investigated and reported separately.

> The main data source will be the WRA's census of waste management facilities, but

> the WRA should supplement the facilities' data with information from the places where the wastes are produced.

Waste management industry

13.12 The waste management industry, and particularly recycling, also should be separated from the industrial and commercial survey: waste may otherwise be double counted[140].

---

[139] EPA90 s75(6) and s75(7), and the Controlled Waste Regulations 1992

[140] Recycling presents additional problems because waste inputs are only 'intended for recycling'. The recycled material will be only one of the outputs from e.g. an MRF: rejected material will continue as waste.

Minerals and agriculture

13.13 The primary industry sectors (minerals and agriculture) should be dealt with separately from the industrial and commercial survey.

Creating the separate groups

13.14 If the WRA is combining the survey with its visits to special-waste producers, group A should be generated first. The source is the WRA's consignment note records, **not** the allfirms file.

13.15 These group A firms must be prevented from re-appearing in groups B to E.

- a The allfirms file should be systematically searched to identify the group A firms. Each record should be flagged (e.g. by entering an A in a 2-character field called GROUP).

- b The flagged records can be copied to a new data file for group A, if that is how the database is designed. But see 13.5 above.

13.16 To create groups B, C, D and E, the companies **not** flagged A in the allfirms file should be indexed in SIC(92) order.

- a Those companies whose SIC codes equate with commercial waste (see Appendix B, table 2) should be flagged CE.

- b The companies still without flags (that is, with a zero-length string in the GROUP field) are the industrial companies (excluding special waste producers). They should now be flagged BD.

13.17 Groups B and D are now extracted from group BD.

- a The BD records should be reordered by number of employees.

- b The top one-fifth (20% of the number of companies[141], but about 80% of the employees) should be re-flagged B. The remaining records are re-flagged D.

13.18 A similar procedure separates group C from group E.

Sample sizes

13.19 Slightly different approaches should be adopted for each group, because a higher level of accuracy is required for the larger producers.

Groups A, B and C

13.20 All firms listed in groups A, B and C should be contacted: but see paragraphs 13.24 to 13.28 for how to do it.

Provide weight data

13.21 The WRA should obtain weights for **all** the waste streams at the firms in groups A, B and C.

- When the WRA first approaches these firms, it should

    > let them know that it will be asking for weight data; and

    > suggest that, if they do not already have these data, they should arrange to get them, either from their own records or from their disposal contractor.

Groups D and E

13.22 For groups D and E, only a sample of firms need be contacted.

---

[141] This number should encompass the whole population being studied. Because in most cases the listing of firms will be a sample, the total from the Census of Employment should be used.

Data collection methods

13.23 Where part of a group is being studied, the WRA should make use of existing survey data (preferably considered across the region) to calculate the sample sizes.

13.24 Some data collection methods need more staff resources than others. The WRA will wish to take account of these resource implications when designing the survey.

13.25 Data collection methods are considered here from the perspective of sampling only. Chapter 14 discusses data collection by personal visit. Chapter 15 discusses data collection by questionnaire.

The visit compared with the postal survey

13.26 The postal survey and the personal visit are the two main survey methods. Table 13.2 compares them.

Table 13.2

Comparison of the two main survey methods

key
✓ = outcome on this facet favours this method

facet	Personal visit survey		Postal questionnaire survey	
Response rate	High (generally 100%)	✓	Low (generally 30%)	
Reliability and accuracy of data	Higher (through supervision)	✓	Lower	
Cost per location surveyed	High		Low	✓
Coverage (function of unit cost)	Low		High	✓
Effect of geographical dispersion on cost	Strong		Weak	✓
Face-to-face contact	Yes	✓	No	
Opportunity for advisory work	Yes	✓	No	

13.27 Visiting each firm in person allows the WRA's staff to

> advise on waste minimisation

> survey waste management practices

> increase the accuracy of the data

but may not always be practicable. The postal survey may be less accurate than the visit because the respondent is not supervised; but it should not be discounted, since

- even data collected during a personal visit will contain errors
- a larger sample will reduce statistical error
- for most planning purposes, less reliable numbers – if known to be so – are better than no data at all.

13.28 Commercial waste is less varied than industrial waste: given an adequate sample, a postal survey of commercial companies can produce useful information. Even so, the WRA should aim to visit the largest commercial companies (largest, that is, as measured by number of employees – group C).

> If a visit is impossible, the WRA may use a postal questionnaire.

Resources available	13.29 A full survey will only be done once every three years. But the WRA, in deciding how many people it needs, should assume that the investigation, and its other waste management planning tasks, are on an annual cycle.
Dealing with data in financial and calendar years	13.30 Time-based data should always use the quarter as the unit. They can then be aggregated into either financial or calendar years. Where this cannot be done, the financial year should be used. As far as possible, quarterly reporting periods should be aligned with accounting periods used in business.
Staff days available	13.31 The number of staff days available during the survey period should first be reckoned.
Effect of deadlines	13.32 Any deadline that cuts the time for the survey will almost certainly increase the survey's cost or decrease its scope and usefulness.
Number of visits achievable	13.33 The number of visits achievable is calculated from the number of staff days actually available in the period of the survey, multiplied by the number of visits per day.
Rural rate different from urban	13.34 When no reliable operational norms can be derived from past surveys, the WRA's staff may assume that each member of staff engaged full time on the visits will make

- an average of 4 visits a day (including preparation time) in urban and suburban areas, but
- in rural areas, where travelling time per visit is likely to be longer, an average of 3 visits a day (including preparation time).

13.35   If the WRA has a mixture of urban and rural sites to visit, it should make an approximation for the split – by numbers of firms or sites – between the urban set and the rural set.

Deciding the approach

13.36   The main options are

> visit all the companies in the sample

> use a postal questionnaire for the smaller commercial waste producers, but visit everybody else

> visit groups A, B and C; survey groups D and E by postal questionnaire

> visit only groups A and B; survey groups C, D and E by postal questionnaire.

13.37   The WRA may wish

> to use a combination of visits and postal questionnaires for a particular group, or

> to visit less than the whole population in groups B and C.

These options are particularly worth considering in the intervening two years between full surveys.

13.38   In making the choice, the primary emphasis should be on the quality of the information that can be produced, rather than the number of firms surveyed.

Number of visits required

Calculating the sample size

13.39 Together with other authorities in the region, and with statistical advice, the WRA should use information obtained from previous surveys or a pilot survey to

> estimate the standard deviation or the variance for each SIC class

> calculate from the standard deviation or variance the sample size in each SIC class that will produce the required accuracy.

13.40 Where WRAs have little information on the class to calculate the required sample size, they may either

> aggregate the figures into SIC Divisions

> calculate the standard deviation or variance for each Division

> use the Divisional figure deviation as the surrogate for the class figure for every class within the Division.

or

> carry out a pilot study to establish an estimate for the required variance

or

> explore whether the National waste database contains an estimate for the class(es) required.

13.41 To calculate the total number of visits in Group D (which contains 80% of the industrial firms), the WRA sums the minimum sample size for each class. Similarly for the visits in Group E.

13.42 Where Groups A, B and C are a 100% sample, the number of visits in these groups is the same as the total number of firms.

13.43 The total number of visits is then the sum of the numbers for Groups A to E.

Shortfall in resources

13.44 If the calculated number of visits will cost more than the budget for the survey, the WRA should increase its resourcing of the survey.

13.45 If the WRA cannot find the full shortfall, the WRA should consider visiting only a sample of the large commercial premises.

> The sample must at least equal the minimum size calculated from the accuracy requirement and the standard deviation.

13.46 If further savings are needed, the WRA should consider using postal questionnaires,

first for Group E (the smaller commercial premises)

then additionally for a proportion of the large commercial premises (a proportion of Group C).

13.47 If the WRA cannot afford to achieve the minimum sample sizes, sampling error will make the survey data largely valueless.

Assessing the time needed for administering postal questionnaires

13.48   Even when the WRA is using a postal questionnaire for economy, staff time must still be allocated to administering it: that is, to issuing questionnaires, chasing up firms who have not responded, and clarifying the responses of those who have.

13.49   If the WRA has time and cost data for these activities from previous surveys, it should use them with the guidance that now follows. If not, the times and proportions given here will allow the WRA to estimate the resource requirements.

13.50   The reckoning of the number of questionnaires to issue depends on an assumption about the likely response rate.

> In general, the response rate for postal questionnaires without any follow-up is about 30%.

> Hence the minimum sample size for each class should be multiplied by 3.3 to give the number of postal questionnaires to be issued in that class.[142]

> Totalling these class results gives the number of questionnaires to be sent out at the start of the postal survey.

13.51   As for follow up, the WRA might reasonably assume that about four-fifths (i.e. about 80%) of the responses will need further action.

> Typically, this will be a 'phone call or fax to clarify answers or complete missing sections.

13.52   To take a hypothetical example (i.e. **the assumed sample size and the constants are hypothetical**):

	activity	Class x	Class 15.11	Class y
a	minimum sample in the Class		10	
b	number of postal questionnaires issued to the Class [3.3*a qu'aires]		33	
c	time to number and envelope questionnaires @ 0.5 minute per questionnaire [0.5*b minutes]		16.5'	
d	minimum time to follow up [0.4*a*3 minutes]: see paragraph 13.53		12'	
e	time to clarify responses [0.8*a*10 minutes]		80'	
f	total time on administering the postal questionnaire for this Class [c + d + e minutes]		108.5'	

13.53   The WRA might decide that, rather than sending out 33 questionnaires to get 10 responses, it would send out fewer questionnaires but chase up non-responders more vigorously.

> This **extra** chasing-up time would be added into the total time needed.

---

[142] The WRA may be concerned that this technique might bias the results by over-representing firms who are more willing to respond. Results in similar surveys suggest that this does not happen. See for example *Recycling of Commercial Waste in Sheffield* (M.E.L Research report 91/02), p.7: '... there is no evidence that response rates are significantly higher in certain SIC Divisions than in others, or that response is more likely from larger premises than smaller premises (or vice versa)'.

> The unit time per issued questionnaire would probably be higher than with the large sample, since the cost of failure is higher. But the response rate, even after several 'phone calls, is unlikely to be higher than 65-75%.

So this is a possible option, but certainly not an easy option.

Final decisions on the minimum sample size

13.54   If the WRA does not have the resources it would need for a postal survey, it must reconsider its options. The integrity and accuracy of the data must not be sacrificed for expediency.

Selecting firms to survey by personal visit

13.55   The WRA

> decides the number of visits to be made in each SIC Class

> makes a random selection of the firms to be visited.

Size bands

13.56   The WRA begins by taking each SIC Class in turn, sorting the firms by number of employees (descending order), and dividing the sorted list into three size bands.

13.57   The banding depends on the listing of firms the WRA has decided to use.

> If the WRA is using the Census of Employment listing, these bands should be 1-24, 25-199 and 200+.

> If the WRA is using the Business Database, the bands should be 1-19, 20-199 and 200+.

> If the WRA is using listings other than these, it should create size bands similar to these.

13.58   For each SIC Class, the WRA then takes the number of visits assigned to the Class, and divides it across the three size bands proportionally to the numbers of firms in each.

> Where the number does not divide exactly, the additional visit(s) should be allocated to the size band containing the largest number of firms.

Selection of firms

13.59   Once the WRA has settled the number of visits in each size band in the Class, it selects the sample of firms randomly. This may be done by either systematic random sampling or true random sampling.

The sampling interval

13.60   The WRA calculates the sampling interval (ie whether to select every 10th firm, or every 50th firm or every 111th firm) by dividing the number of firms in the size band by the minimum sample size for that size band. Results that are not whole numbers are **rounded down** rather than rounded up.

13.61   For example: assume, hypothetically, that – in the WRA's area – SIC subsection DA class 15.11 (Production and preserving of meat) has 86 firms in the 1-24 employee band, and that the required sample size is 10. Then

> $86 \div 10 = 8.6$

> 8.6 is rounded down to 8.

The sampling interval is 8.

Making the selection

13.62 The WRA selects the sample of firms. The hypothetical case (paragraph 13.61) provides a sampling interval of 8, and a required sample of 10. The procedure for selecting the **systematic** random sample is A, B, and C below; the procedure for selecting the **true** random sample is A, B and D.

SYSTEMATIC RANDOM SAMPLE

A From the starting point[143] the random number table delivers 95 97, 37 99, 05 79, 55 85, 42 28, 4 19, and 49 31.

B 95 and 97 are outside the range 1-86: so the sample starts with the 37th firm.

C The WRA, noting that the required sampling interval is 8, flags records 37, 45, 53, 61, 69, 77, 85, 7, 15 and 23.

TRUE RANDOM SAMPLE

A From the starting point the random number table delivers 95 97, 37 99, 05 79, 55 85, 42 28, 4 19, and 49 31.

B Again, 95 and 97 are outside the range 1-86: so the sample starts with the 37th firm.

D The WRA, proceeding through the sequence of numbers from the random number table, flags records 37, 05, 79, 55, 85, 42, 28, 4, 19 and 49.

13.63 This procedure is repeated for every size band in every SIC Class.

Selecting firms to be surveyed by postal questionnaire

13.64 The postal survey is not the method of choice: see paragraph 13.53 above. But it may sometimes be unavoidable.

Number of firms

13.65 The WRA's staff should get their data on the number of firms in the area from the Census of Employment, **not** from the listing of firms.

> This is because the listing of firms may only include a sample of those employing less than 25 staff: but the number of questionnaires must be calculated from the required proportion of the **total** population.

13.66 If x represents the minimum sample size, questionnaires must be sent to at least $3.3x$ firms.

If sample size exceeds population

13.67 In most years the Census of Employment lists only a sample of firms (see paragraph 13.3). The WRA's minimum sample size – the $3.3x$ in paragraph 13.66 – could conceivably be larger than the number of firms in the Census of Employment list (or in lists derived from it).

13.68 If this happens, the WRA has little room for manoeuvre. The WRA should

---

[143] The starting point should either be chosen at random, or be the next number in sequence after the last stopping point.

- send a questionnaire to every company on the list
- use the time saved (by having to dispatch fewer questionnaires) to chase up non-respondents – with a reasonable amount of follow up work this should increase the response rate to 50% or more.

Selecting firms to survey by post

13.69 In selecting firms to survey by post, the WRA should follow the same procedure as for personal visits: see paragraphs 13.55 to 13.63.

13.70 As with selecting firms for personal visits, the WRA should

> group firms on the list by SIC Class, and group each Class in three size bands

> use random sampling (either systematic or true) to select – within each band in each SIC – the firms who will be sent questionnaires.

Creating the final listings of firms to be surveyed

13.71 The firm's record in the allfirms file should be flagged to show whether it is to

> receive a visit OR

> receive a questionnaire, OR

> be excluded from the survey.

13.72 When this has been done, the WRA's staff may wish to

> create two new database files – one for visits, the other for postal questionnaires

> copy (not move) the visit records from the allfirms file to the visit file

> copy (not move) the postal-questionnaire records from the allfirms file to the postal questionnaires file

> archive the allfirms file for use in future surveys.

Creating the two new files may reduce the chance of mistakes – visiting unselected firms, for example.

Chapter 14 Guidelines for survey by personal visit

14.1 This chapter gives guidance to WRA staff on planning, preparing for and carrying out personal visits to selected firms. It suggests procedures for

> planning visits efficiently

> contacting the companies and confirming the arrangements

> using the visits to achieve the survey's objectives.

It also considers

> the design of a form to record data during visits

> what data should be collected, and why.

Approaching firms to be visited

14.2 Visits to firms in the same neighbourhood should be programmed for the same day.

Group the firms by location

14.3 The WRA staff should

> call up the lists of firms to be visited, indexing them in postcode order

> mark them off (probably manually, to start with) into daily groups

> assign a survey officer and a date to each group

> enter the survey officer's name (or initials or other code) and the visit date in the fields SURVEYBY and VISITDATE

> after indexing on these fields, print them out as work lists for each survey officer.

Appointment

14.4 At least 3 weeks before the proposed visit, the survey officer should 'phone each firm on his work list to

> identify a contact

> confirm the name and address of the company

> make an appointment.

Contacts at firms producing special waste

14.5 For a contact at a firm producing **special** waste, the survey officer should check the WRA's special waste database.

DATA PROTECTION

14.6 Here, as elsewhere in this guidance, the purposes the WRA has declared to the data protection registrar may constrain its use of its database for new purposes.

Confirm by letter

14.7 At least 2 weeks before the date of the visit, the survey officer should write to the firm to

> confirm the date and time of the visit

- > explain the purpose of the survey
- > say how long the visit is likely to take
- > indicate what agenda the WRA has in mind.

14.8 The letter should also

- > indicate that better information about wastes, besides improving waste collection and disposal, will benefit the waste producers themselves; and that the WRA will feed summary data back to the company: see chapter 7, paragraphs 7.38 to 7.40

- > explain that the survey officer will offer advice on waste management and waste minimisation, if the firm would find that helpful

- > stress that a sufficiently senior member of the company's staff should be present during the visit, particularly during the discussion of waste minimisation (as decisions about waste minimisation are often made at board level)

- > emphasise the importance of weighing waste

- > make it clear that the survey officer will wish to
  - > examine the company's waste records, including those that show
    - > the quantity of waste removed over the last year
  - > see all the operations that generate waste.

Designing the data forms

14.9 Survey staff visiting firms might use a three-part form (an example form is printed at Appendix C):

part 1 data on the firm

part 2 data on each of the waste streams separately

part 3 data on waste management practice at the firm.

14.10 Table 14.1 below indicates the dataset that part 2 should include for each waste stream, and why. This dataset should be seen as the minimum. If the WRA has sound reasons, it may collect additional information. But if it does so,

- the WRA must also ensure that collecting the additional information makes **no** significant **net** addition to the burden on business –

- that is, unless the demands are small, business's marginal resource cost for collecting the additional data should be matched by the increased value of the WRA advice that results from analysing the additional data.

Table 14.1

Data to be collected about each type of waste, and the reason for its collection

serial	data to be collected	reason for collection
1	Waste description and waste code	To identify and encode the type of waste
2	Process producing the waste	To identify the process generating the waste, thus clarifying the causes of waste generation.
3	Component materials of the waste, with %	To identify the materials making up the waste: this is particularly important for general waste which may contain recyclables or packaging
4	Significant contaminants in or on the waste	To be able to advise on the suitability of disposal routes for the waste
5	Whether the waste is mixed in the container	To assist in deciding whether the waste is potentially available for reclamation
6	Weight of the waste (where known)	To provide an accurate measure of the quantity of the waste
7	Form of the waste	To assist planning for different types of waste management operation
8	Whether the waste is routinely produced i.e. directly related to manufacturing process or service	To avoid error when scaling up from the sample to the whole population. The typical company by definition probably produces some wastes routinely; but **not** all companies produce the one-off wastes (such as waste from stripping asbestos).
9	The quantity of waste that is special waste	Special waste needs special care in its management. (The estimate may be validated through use of the consignment note records.)
10	The quantity of waste that is packaging waste and its type	To show progress towards targets (those within the Packaging Directive, and those suggested by the Producer Responsibility Group for packaging)
11	Type of container	To estimate waste volume as a surrogate for weight (where weight data are not available, or to improve conversion factors), and to strengthen advice on waste management
12	Number of containers	To estimate waste volume as a surrogate for weight
13	Number of containers collected or emptied per annum	["]
14	Size of the container	To estimate waste volume as a surrogate for weight
15	How full the container usually is when collected or emptied	["]
16	If the waste is compacted, the compaction ratio	["]
17	Name of the waste carrier	As a source of information if the respondent is unclear, or as a cross-check
18	Carrier's registration number	To assist in associating carrier and waste
19	Method of waste transport	As one component in an aggregated view of waste movements within the area
20	Method of disposal or treatment	To help in targeting waste reclamation and reduction advice, and in planning for new waste disposal sites
21	Name of disposal or treatment site	For cross checking; and to pass to the receiving WRA if the waste is not dealt with in the home WRA's area
22	Location of the disposal or treatment site	To assess the area's progress towards conformity with the proximity principle and with regional self-sufficiency
23	Expected changes in quantity of the waste over the next few years	To provide a basis for forecasting the total of the waste stream over the next few years.

The visit

Preparation

14.11 The survey officer should prepare for each visit by

> reviewing what he knows about the firm's operations, or those of firms in similar businesses

> making a provisional list of the waste streams those operations are likely to produce. Besides preparing the mind, this list is used during the visit: see paragraph 14.26 below

> considering opportunities for minimisation, recycling and recovery of waste.

When to complete the parts of the visit form

14.12 The survey officer should complete

- **part 1** of the visit form during the preliminary discussion in the respondent's office

- **part 2** when inspecting the operational area of the site (one set of answers for each waste stream)
- **part 3** on return to the respondent's office following the inspection.

The opening discussion: part 1 of the visit form

14.13   The visit should start with an examination of the waste producer's paperwork about waste. Although the paperwork will usually **not** serve to quantify waste production, it will provide an overview of the different waste streams. It may also alert the survey officer to areas of difficulty.

14.14   The survey officer completes part 1 of the form at this point.

Site inspection: part 2 of the visit form

14.15   The survey officer inspects the operational areas of the site to identify and quantify

> the waste streams

> the processes that produce them.

The survey officer completes one copy of part 2 for each waste stream. The set of these forms constitutes part 2 of the visit form.

Materials reclaimed or re-used on site

14.16   The survey officer must try to cover **all** the waste streams the company produces. He should include

> wastes for reclamation

> wastes re-used on-site (except where this is as an integral part of a process).

14.17   The producer may not see material used on site as waste: it is not discarded now, so the producer assumes it never will be. But a change in the process technology could make it waste and entail its removal.

14.18   From the WRA's point of view, a record of re-use is worthwhile because it gives a more accurate picture of the total beneficial use of waste.

14.19   The survey officer should record re-use by indicating under *Method of treatment and disposal* that the waste is reused wholly on site.

Quantifying waste streams

14.20   Once identified, the waste stream needs to be quantified as accurately as possible.

ACTUAL WEIGHT MORE SATISFACTORY THAN CALCULATED WEIGHT

14.21   Some companies may weigh their waste as or before it is removed, or may have invoices showing the weight at the disposal or treatment site.

- These actual weights should be used in preference to weights calculated by conversion from volumes.

14.22   Volume conversion is sometimes unavoidable: see chapter 16, paragraphs 16.20 to 16.40, for the technique. The WRA's staff will then need to use

> the volume data listed at serials 11 to 16 of table 14.1, together with

> the standard factors (when these are available) for converting volume to weight (see paragraph 16.34).

Changes in the waste streams

14.23 The survey officer asks the respondent to indicate any foreseeable changes in waste generation over the next few years.

14.24 Sometimes the respondent can foresee, and divulge, a major business change: for example, changing the process machinery, or bringing in a relocated process line from elsewhere.

14.25 The survey officer should record the change itself in part 3 of the form; and should show in part 2 of the form the respondent's estimates of the percentage change it will cause in each waste stream.

- If the respondent is **not** able to estimate, the survey officer should substitute his own approximations, annotating the form to show that is what they are.

Some waste streams often overlooked

14.26 Before leaving the operational area of the site, the survey officer should

> cross off from his provisional list (see paragraph 14.11) the waste streams he has observed

> ask about those that remain on the list.

14.27 Respondents can easily overlook

> waste from the site office

> waste from the canteen

> pallets

> drummed waste

> loose waste

> waste taken to a waste management facility by the firm itself

> scrap metal

> waste oil

> other materials not seen as waste because the firm recycles them itself.

Ending the visit: part 3 of the visit form

14.28 Once he has finished the site inspection, and before he leaves the site, the survey officer should

> ask the respondent to find out any missing data

> fix a day and time for the survey officer to 'phone the respondent for these data

> complete part 3 of the form.

After the visit

Recording visits to special waste producers in groups B,C,D and E

14.29 Where Group A does not include all producers of special waste, visits to firms in the other groups will almost certainly include some to special waste producers. Where these are visited as part of the survey, the visit should be recorded on the consignment note records as a periodic inspection.

Feedback report

14.30 Wherever possible – and certainly where there are recommendations made, or to be made, that significantly affect the management of waste at a site – the WRA should send to the waste producer a summary of the facts found on the visit and any associated recommendations or details of information requested by the producer during the visit (see paragraph 7.38 to 7.40).

Chapter 15 Guidelines for postal surveys

General administration

15.1   The first stage of the postal survey is to

> mark the database records of companies who will be sent the questionnaire

> copy the records into a separate file, **if** that is the WRA's preferred way of working.

15.2   The record structure should allow the survey staff to print address labels that carry the **producer reference number** as well as the postal address.

> If the WRA already knows, reliably, who its contact in the respondent firm should be, the label should be addressed to the contact by name.

> Otherwise the WRA should 'phone the firm first to find out the name of the right contact, then address the label to the contact by name.

15.3   A letter should accompany the questionnaire. It will resemble the pre-visit letter described in chapter 14, paragraphs 14.7 to 14.8. It should

> explain the purpose of the survey

> indicate the survey's statutory legitimacy

> indicate the benefits to waste producers themselves

> offer a visit to give guidance on waste management and waste minimisation, **if** the firm would find it useful and requests it

> undertake, in any case, to feed back to the firm the summary data from the survey

> give a name and number for queries.

15.4   The WRA should send a freepost[145] return envelope with each questionnaire.

> The WRA might send a pre-addressed envelope, leaving the firm to pay the postage. This is a false economy if it significantly reduces the response rate, which it probably will.

15.5   The record structure should also allow the WRA to

> check which firms have and have not returned the questionnaire

> output address labels or 'phone lists so that non-respondents can be reminded.

---

[145] Freepost accounts can be set up by applying to the Post Office.

Questionnaire content and design

15.6 The questionnaire is a substitute for a personal visit. It may not be a wholly satisfactory substitute; but within reason it should be designed to collect the same data as a survey officer would collect on a successful visit.

- The data items are listed in table 14.1: the questionnaire designer should consult that table.

15.7 The questions should be straightforward and self-explanatory. The respondent will fill in the form more readily, and probably more accurately, if the form designer

> breaks down the data into component items shown as they are likely to occur in the business

> uses tick boxes to simplify the answers and lead the respondent through the form

> avoids long instructions.

15.8 Example questionnaires are set out in Appendices D and E.

Mailing

15.9 The WRA staff should mark each questionnaire with its producer reference number before sending it out.

- The WRA can then file the completed questionnaires in the same sequence as the database records. This makes it easier to retrieve the hard copy.

15.10 The most efficient way of linking questionnaire to envelope is probably to

> number each of the questionnaires with the producer reference number: this can be done

   a with a pen, or

   b by printing the numbers from the database direct onto the front sheets, or

   c by printing the numbers from the database onto labels, and sticking these to the front sheets (in step with labelling the envelopes), or

   d by printing one comprehensive label per respondent, sticking it to the front sheet, and using window envelopes.

> (With methods a and b) keep the addressed envelopes (whose labels carry the producer reference numbers) and the questionnaires in producer-number order.

> (With methods a, b and c) make regular checks when putting the questionnaires into the envelopes to ensure that envelopes and forms are correctly matched.

15.11 The documents should be put in the envelope in a sequence that will influence the recipient to respond. Probably the recipient should see

first the letter

second the questionnaire

finally the return envelope.

Large mailings 15.12 If the mailing is large, it may be more efficient to

> contract it out (specialists exist in most areas), or
> use temporary staff.

15.13 If the WRA uses either of these methods, it must ensure that the people doing the work

> have instructions that explain how to match the questionnaire number with the number on the envelope label
> understand why this matching is important
> preferably have an incentive not to get it wrong.[146]

Staggering the responses 15.14 If all the questionnaires are posted on the same day, more than half of all the responses that can be expected will arrive in the following three weeks. During these three weeks the survey officers will probably spend much of their time following up the responses: they are unlikely to be available for other work – most obviously, survey visits.

- The WRA should therefore plan the programme of survey visits so that it will reflect the numbers of forms likely to come back over the six weeks following dispatch of the questionnaires.

15.15 The alternative is to post the questionnaires in, say, six batches over six weeks. This flattens the peak: about 7% would come back each week. The follow-up work is still the same in total, but can now be carried on in parallel with some survey visits.

Dealing with returned questionnaires 15.16 When a questionnaire is returned, the WRA staff

> write the date of receipt on the front of the questionnaire
> enter the date of receipt in the database record.

Obvious gaps 15.17 The WRA staff **promptly**

> check the form for obvious gaps (e.g. essential boxes left blank), and
> mark all such gaps.

Essential data items 15.18 The most important data items are

> waste type
> waste weight (or the volume surrogates for weight)
> number of employees in the firm
> the firm's SIC codes to sub-class level.

The importance of the first two data items is obvious enough: that of the latter two, less so. Yet much of the estimation depends on employee count and the SIC code.

---

[146] If the work is done under contract, the contract may include a penalty for unacceptably poor performance – which the contract will have to define.

Less obvious gaps

15.19 The survey officer then **promptly** checks the form for less obvious gaps: for example, waste streams unaccountably missing.

Action

15.20 The survey officer 'phones (or faxes) the contact at the respondent firm within one day of receipt to get the gaps filled in.

- Delay seriously erodes the chances of getting meaningful data.
- The survey officer should **not** delegate contact with the firm **unless** the only gaps are those found by the support staff. Approaches from several people will confuse and annoy the respondents.

Increasing the response

15.21 If after four weeks the response is less than about 75% of what is statistically essential, the WRA should consider chasing up firms who have not replied. This may be labour intensive.

> If the WRA has chosen to send questionnaires to a large sample (see paragraphs 13.50 to 13.52 above), it should not be faced with a lower response than it needs.

15.22 If the WRA decides to chase up the non-respondents, it should in principle contact all of them. In practice this may well be unrealistic. Contacting a selection of firms, however, could bias the results.

- The WRA staff can avoid bias by further selecting non-respondents **at random** from each SIC Class.

15.23 Often the non-responding firm will have lost or thrown away its first copy of the questionnaire, and – if persuaded to respond – will need another copy.

- The WRA should anticipate these additional demands for questionnaires, letters and return envelopes by adding an – approximately calculated – number to the original print run (after taking cost advice from reprographics specialists).

> The average cost of extra copies on the margin of the original print run will probably be low: the average cost of copies in a small separate print run would be substantially greater.

Storing returned questionnaires: archiving data

15.24 The WRA should store the returned questionnaires in producer number order.

- The store should be secure, since the data in the questionnaires may aggregate to commercially sensitive information.

15.25 In archiving electronic data, the WRA will take account of its obligations under the Data Protection Act 1984.

Retention periods

15.26 The WRA will wish to keep the returned questionnaires from year i until at least the time when it begins to plan year $(i + 3)$'s survey – i.e. the return to the equivalent year in the next 3-year cycle.

15.27 In principle the WRA should retain the whole run of questionnaires until every firm has been surveyed at least once. But since some small firms may not be visited more than once every 10 years, the WRA may decide that the storage and administration costs are excessive.

- A realistic compromise is probably to assume a 3-year retention period, but to review this after the first 3 years.

Destruction

15.28 Once a batch of returned questionnaires ceases to justify its storage and administration costs, the WRA will wish to dispose of it. Even after the batch is marked for disposal, it should continue to be treated as commercially sensitive, and disposed of accordingly.

- The WRA should dispose of returned questionnaires by either shredding or burning.

Chapter 16 Data management, checking and validation

Computer is better for managing large datasets

16.1 The WRA will normally use computers for managing large datasets. This reduces

> time spent on collating and analysing data

> time spent on formatting data for output

> the potential for human error.

16.2 The WRA will wish to ensure that proper data handling procedures are used.

- In particular, data should be backed up regularly and frequently.

Spreadsheet or database?

16.3 Survey data may, in principle, be managed with either a database package or a spreadsheet. The spreadsheet at first seems easier; but

> data validation by software is less satisfactory on a spreadsheet

> the two dimensions of the spreadsheet layout will cramp the data, which are *n*-dimensional.

A database should preferably be used to handle survey data, unless – improbably – the WRA has nobody with database skills **but** somebody with advanced skills in spreadsheets.

16.4 The waste survey is **not** an occasion for piloting unknown packages.

> If the WRA already knows and uses one of the packages designed to handle survey data, well and good. If not, the WRA's usual database is the right choice.

Database design

Files

16.5 Raw survey data should be held in a data-entry file or files.

- Data in data-entry files should not be manipulated, except by being copied to **separate** database files for subsequent manipulation.

16.6 The package must be able to output from all database files (including the data-entry files) in a standard computer-readable format[147]. This allows the convenient transmission of the data to other systems.

Fields

16.7 Each data item in the questionnaire should correspond to a field in the database. (The reverse will probably not be true.)

16.8 The field list for the data-entry files should include a visit date field and a survey reference field. They should be completed in every record: coded entries are of course allowable.

File relationships

16.9 The database will be a set of one-to-many relationships, since any one firm typically produces more than one waste.[148]

---

[147] most obviously ASCII, CSV, or dBase .DBF

[148] The state of the real world is of course many-to-many: lots of waste producers contributing to lots of waste streams. This is not a model the normal database can cope with.

- A flat-file database can represent one-to-many relationships, but does it wastefully.

- Hence the WRA's database design will probably use several related files[149], with the *producer reference number* as the link.

16.10   The Department of the Environment is considering the development of a standard database structure for use by WRAs and the Environment Agency.

Data coding

16.11   All questionnaires, and the database, should have the data items in the same sequence. This makes data entry easier.

16.12   Encoding the survey data facilitates subsequent analysis. To enter data in a coded format,

> **either**   the codes may be written before data entry
>
>> **either** on the returned questionnaires
>>
>> **or** on a separate sheet
>
> **or**   the data may be coded as they are entered.

- Details of the type of codes required (known as a coding frame) and an example coding form are given in Appendix F.

16.13   Some software packages allow data entry direct from the questionnaire. The package, suitably set up, will

> encode the data
>
> reject entries outside a specified range and type
>
> validate the data against pre-determined relationships between fields.

Data checking and validation

16.14   As a check, the WRA should

> make its own approximations for the weight of each major waste stream
>
> use the database to sum the data for the waste arisings in each major waste stream
>
> compare the two
>
> analyse the discrepancies.

16.15   If the data were not validated during entry, this must be done before the WRA tries to get useful work out of the system.

- Range checks should be set up for each field.

16.16   Where the range check shows an out-of-range entry, the WRA should

> check the coding sheet, and correct it if necessary

---

[149] probably **not**, however, *relational* in the strict sense defined by Codd

Possible errors in the essential data items

> (if the data coding sheet is wrong) check the entry on the questionnaire itself

> correct the data item in the database file

> decide whether the error is random or systemic

> take action to correct systemic errors.

16.17 The most important data items are

> waste type

> waste weight (or the volume surrogates for weight)

> number of employees in the firm

> the firm's SIC codes to sub-class level.

16.18 If validation checks show up actual or possible errors in these data items, the WRA should renew contact with the firm to correct the errors.

Deriving weight of waste from volume of waste

16.19 Some firms cannot provide weight data for some wastes. Immediately after data entry, therefore, some database records have a zero in the WEIGHT... fields. But, if the survey has gone well, the record shows volume data instead. Volume data can be converted to or provide an approximation of weight.

16.20 Database commands for this kind of manoeuvre are specific to the database package. The examples that follow use xBase as a pseudocode.

16.21 The prudent database user will manipulate only **copies** of the database files. The original files will be kept out of reach, and set to read-only.

Load volumes

Natural volume

16.22 For any one waste type, the natural volume of a load depends on

1 the container size

2 the compaction ratio[150]

3 how full the container is.

Thus a 9 cubic yards container, 90% full and compacted in a 2:1 ratio, has a natural load volume of (9 x 0.9 x 2 =) 16.2 cubic yards.

16.23 For conversions to weight, the natural volume is better than the compacted volume because

- the standard density factors will be for **un**compacted wastes (see paragraphs 16.33–16.35).

16.24 The xBase command for getting these volume data from their fields in a record, multiplying them, and putting the result into a record field for imperial volume, is

. replace VolImp with CntrSize * Compact * HowFull

---

[150] The compaction ratio is the natural volume of the waste *divided by* its volume in a container where it has been mechanically compressed.

Metric volume

16.25   In practice, the user will need to know **metric** volume: most of the standard density factors – see paragraph – will be in metric units.

16.26   The machine converts the volume to metric if a field called, say, SIZEUNIT = 1: this tells it that the container size is in cubic yards. The conversion command, on its own, would be

. replace VOLMET with (VOLIMP * 0.76) for SIZEUNIT = 1

If the container size were already in cubic metres, SIZEUNIT would be set to zero. Other units of volume could be accommodated in a similar way.

16.27   The conversion factors – for example, the 0.76 used in the command in paragraph 16.26 above – would normally be held in a variable. They are shown here as absolute values to clarify the examples.

Annual volume

16.28   Finally, the user wants to know the **annual** volume, rather than the volume of one load: the assumption is that the one-load volume is typical. The machine

> gets the number of loads per year from the record field NUMPERA

> multiplies the one-load metric volume by the number of loads per year

> leaves the result in a record field called VOLPERA.

The xBase command is

. replace VOLPERA with VOLMET * NUMPERA

Single command

16.29   For clarity, this account has broken down the procedure into steps. In practice (and with practice) the machine can be told to do the volume conversion, on every record that needs it, in a single operation.

16.30   The following xBase command calculates annual volume in every record where the unit for container size (CNTRSIZE) is cubic **metres**.

. replace VOLPERA with CNTRSIZE * COMPACT * HOWFULL [* 1.0]

* NUMPERA for SIZEUNIT = 0 .and. WEIGHTPERA = 0

The machine has to be told **not** to do any of this if the record already contains a figure for weight: hence the condition that WEIGHTPERA must be zero.

16.31   If other volume units need to be catered for, the user issues a similar command, but with other values for SIZEUNIT and the conversion factor.

> Thus for CNTRSIZE values in cubic **yards**, SIZEUNIT would have been set to 1, and the conversion factor (cubic yards to cubic metres) is 0.76.

> The command for calculating annual volume becomes

. replace VOLPERA with CNTRSIZE * COMPACT * HOWFULL * 0.76

* NUMPERA for SIZEUNIT = 1 .and. WEIGHTPERA = 0

Check	16.32 Before continuing, the database user will check that the manoeuvre has worked.
Converting cubic metres to tonnes of waste	16.33 Standard density factors will be the subject of consultation by DoE as part of its work on the development of the national waste database.
Density factors	16.34 The standard density factors will be mostly in **metric** units. The user therefore has three options

        1   record all volume data in metric units

        2   provide additional density factors by conversion

        3   convert non-metric data to metric.

The choice should have been made at project design stage. The examples in this chapter have assumed option 3.

Linking waste type and density factor	16.35 In a separate database file the user links each waste type with its density factor. A small program then reads the waste type in each record, and inserts the density factor that matches the waste type.

> Or, less elegantly, the user can do this with (in xBase) a series of replace commands.

Calculating annual equivalent	16.36 A further replace command then goes to each record where WEIGHTPERA is 0. It multiplies annual waste volume (VOLPERA) by the density factor, and puts the result in WEIGHTPERA:

. replace WEIGHTPERA with VOLPERA * DENSITY for WEIGHTPERA = 0

Check	16.37 The database user will make some further manual checks on a sample of the records, using a calculator. This ensures that the procedure has been done correctly.
Using local knowledge to verify results	16.38 Once a weight has been obtained for every waste stream surveyed, the WRA should check a sample of the results. Each result should make sense: that is, should square with the survey team's real-world and local knowledge.
	16.39 Where a figure is unusually large or small, the WRA should check for errors in the coding sheet **and** the questionnaire.

Appendix A

The Controlled Waste Regulations 1992 (SI 1992 No 588): Waste to be treated as industrial waste

The Controlled Waste Regulations 1992 (SI 1992 No 588): Waste to be treated as industrial waste

1. Waste from premises used for maintaining vehicles, vessels or aircraft, not being waste from a private garage to which paragraph 4 of Schedule 1 applies.

2. Waste from a laboratory.

3. (1) Waste from a workshop or similar premises not being a factory within the meaning of section 175 of the Factories Act 1961 because the people working there are not employees or because the work there is not carried on by way of trade or for the purposes of gain.

   (2) In this paragraph, 'workshop' does not include premises at which the principal activities are computer operations or the copying of documents by photographic or lithographic means.

4. Waste from premises occupied by a scientific research association approved by the Secretary of State under section 508 of the Income and Corporation Taxes Act 1988.

5. Waste from dredging operations.

6. Waste arising from tunnelling or from any other excavation.

7. Sewage not falling within a description in regulation 7 which -

   a) is treated, kept or disposed of in or on land, other than by means of a privy, cesspool or septic tank;

   b) is treated, kept or disposed of by means of mobile plant; or

   c) has been removed from a privy or cesspool.

8. Clinical waste other than -

   a) clinical waste from a domestic property, caravan, residential home or from a moored vessel used wholly for the purposes of living accommodation;

   b) waste collected under section 22(3) of the Control of Pollution Act 1974; or

   c) waste collected under sections 89, 92(9) or 93.

9. Waste arising from any aircraft, vehicle or vessel which is not occupied for domestic purposes.

10. Waste which has previously formed part of any aircraft, vehicle or vessel and which is not household waste.

11. Waste removed from land on which it has previously been deposited and any soil with which such waste has been in contact, other than -

    a) waste collected under section 22(3) of the Control of Pollution Act 1974; or

    b) waste collected under sections 89, 92(9) or 93.

12. Leachate from a deposit of waste.

13. Poisonous or noxious waste arising from any of the following processes undertaken on premises used for the purposes of a trade or business –

    a) mixing or selling paints;

    b) sign writing;

    c) laundering or dry cleaning;

    d) developing photographic film or making photographic prints;

    e) selling petrol, diesel fuel, paraffin, kerosene, heating oil or similar substances; or

    f) selling pesticides, herbicides or fungicides.

14. Waste from premises used for the purposes of breeding, boarding, stabling or exhibiting animals.

15. (1) Waste oil, waste solvent or (subject to regulation 7(2)) scrap metal, other than:

    a) waste from a domestic property, caravan or residential home;

    b) waste falling within paragraphs 3 to 6 of Schedule 1.

    (2) In this paragraph -

    'waste oil' means mineral or synthetic oil which is contaminated, spoiled or otherwise unfit for its original purpose; and

    'waste solvent' means solvent which is contaminated, spoiled or otherwise unfit for its original purpose.

16. Waste arising from the discharge by the Secretary of State of his duty under section 89(2).

17. Waste imported into Great Britain.

18. (1) Tank washings or garbage landed in Great Britain.

    (2) In this paragraph -

    'tank washings' has the same meaning as in regulation 2[a] of the Control of Pollution (Landed Ships' Waste) Regulations 1987[b] and:

    'garbage' has the same meaning as in regulation 1(2) of the Merchant Shipping (Reception Facilities for Garbage) Regulations 1988[c].

---

[a] Amended by SI 1989/65

[b] SI 1987/402

[c] SI 1988/2293

Appendix B

SIC codes for industrial and commercial waste producers

Table B1: SIC(92) Divisions to be classed as industrial waste producers

SIC(92) Division	Description
15	Manufacture of food products and beverages
16	Manufacture of tobacco products
17	Manufacture of textiles
18	Manufacture of wearing apparel; dressing and dyeing of fur
19	Tanning and dressing of leather; manufacture of luggage, handbags, saddlery, harness and footwear
20	Manufacture of wood and of products of wood and cork, except furniture; manufacture of articles of straw and plaiting materials
21	Manufacture of pulp, paper and paper products
22	Publishing, printing and reproduction of recorded media
23	Manufacture of coke, refined petroleum products and nuclear fuel
24	Manufacture of chemicals and chemical products
25	Manufacture of rubber and plastic products
26	Manufacture of other non-metallic mineral products
27	Manufacture of basic metals
28	Manufacture of fabricated metal products, except machinery and equipment
29	Manufacture of machinery and equipment not elsewhere classified
30	Manufacture of office machinery and computers
31	Manufacture of electrical machinery and apparatus not elsewhere classified
32	Manufacture of radio, television and communication equipment and apparatus
33	Manufacture of medical, precision and optical instruments, watches and clocks
34	Manufacture of motor vehicles, trailers and semi-trailers
35	Manufacture of other transport equipment
36	Manufacture of furniture; manufacturing not elsewhere classified
40	Electricity, gas, steam and hot water supply
41	Collection, purification and distribution of water
50.2	Maintenance and repair of motor vehicles
60	Land transport; transport via pipelines
61	Water transport
62	Air transport
63 (not 63.3)	Supporting and auxiliary transport activities
64	Post and telecommunications
73	Research and development
85	Health and social work
90	Sewage and refuse disposal, sanitation and similar activities

Table B2: SIC(92) Divisions to be classed as commercial waste producers

SIC(92) Division	Description
50 (not 50.3)	Sale of motor vehicles and sale, maintenance and repair of motorcycles; retail sale of automotive fuel
51	Wholesale trade and commission trade, except of motor vehicles and motorcycles
52	Retail trade, except of motor vehicles and motorcycles; retail of personal and household goods
55	Hotels and restaurants
63.3	Activities of travel agencies, tour operators etc.
64	Post and Telecommunications
65	Financial intermediation, except insurance and pension funding
66	Insurance and pension funding, except compulsory social security
67	Activities auxiliary to financial intermediation
70	Real estate activities
71	Renting of machinery and equipment without operator and of personal and household goods
72	Computer and related activities
74	Other business activities
75	Public administration and defence; compulsory social security
80	Education
85.3	Social work activities
91	Activities of membership organisations not elsewhere classified
92	Recreational, cultural and sporting activities
93	Other service activities

Table B3: SIC(92) Divisions to be treated separately from the survey

SIC(92) Division	Description
01	Agriculture, hunting and related service activities
02	Forestry, logging and related service activities
05	Fishing, operation of fish hatcheries and fish farms; service activities incidental to fishing
10	Mining of coal and lignite; extraction of peat
13	Mining of metal ores
14	Other mining and quarrying
37	Recycling
45	Construction
90	Sewage and refuse disposal, sanitation and similar activities (but excluding those activities relating to household or industrial waste, eg. collection of refuse from litter boxes, sweeping and watering of paths etc.)
95	Private households with employed persons
99	Extra-territorial organisations and bodies

Appendix C

Suggested format for personal visits questionnaire

Survey of Wastes   Form #1

ENVIRONMENTAL PROTECTION ACT 1990 (SECTION 50)

DATE				PRODUCER NUMBER	
	Day	Month	Year	GROUP NUMBER	

INFORMATION ABOUT THE WASTE PRODUCER

Company name

Address

Postcode

Contact name

Telephone number

INFORMATION ABOUT THE HEAD OFFICE (IF APPLICABLE)

Head Office name

Address

Postcode

Contact name

Telephone number

INFORMATION ABOUT THE BUSINESS ACTIVITY

Nature of business

Main production processes
1
2
3
4

INFORMATION ABOUT EMPLOYEES

Number of employees:		
	Full time	
	Part time	
	Total (FTE)	

Survey of Wastes

Form #2

ENVIRONMENTAL PROTECTION ACT 1990 (SECTION 50)

DATE [Day] [Month] [Year]   PRODUCER NUMBER

WASTE STREAM NUMBER

1 Waste Description

2 Process(es)

3 Waste classification

			%	Code	Mixed?
Component 1					
Component 2					
Component 3					
Component 4					
Component 5					

4 Is it weighed on site?   5 (if yes) Quantity per annum   6 Units

7 Form   8 Special?
9 One-off disposal   10 Packaging?

11 Type of Container   12 Number

13 Emptied/Collected   x every   =   Times a year

14 Size   16 How full?   %
15 Units   17 Compaction ratio   :

18 Carrier Name..................................   Reg. No.

19 Method of transport

20 Disposal/Treatment Method

21 Disposal/Treatment site

22 WRA of Disposal/Treatment site

23 Likely changes ....   + or −   %
                                                (circle one)

Appendix D

Example postal questionnaire for small firms

SURVEY OF WASTES

This survey is about the wastes generated by your premises – that's the premises to which this letter and questionnaire have been addressed. You may be a business or another type of employer (public, private or voluntary). You may produce a lot of waste or only small quantities. In either case, please fill in all sections that are appropriate to your organisation.

This questionnaire should be completed by the manager of the premises, or the person responsible for your company's waste. If a landlord is responsible for your company's waste, you should ask them or any contract cleaners any questions that you are not sure of but still complete it for your company only.

The purpose of this survey is to enable us to estimate how much waste of different kinds is generated by different sizes of company in each industry sector. The results will be used to plan for the future to help ensure continuity of waste management outlets for waste from your business. It will also add to information on the national waste database – a database of information on waste arisings that you can access to compare your waste production figures with those of your industrial sector generally. This will help you see whether you might be able to save money easily by producing less waste.

The survey concentrates on weight data wherever these are available because they will provide you and us with a much more accurate measurement of the quantities and therefore the cost.

This is part of an official survey on behalf of [WRA]. If you have any queries, please contact [xxx] on [telephone number]. Thank you for your co-operation.

Remember: the cost of waste is not only what you pay for its removal, it is the cost of the materials that are lost.

Please be assured that all information provided will be treated in the strictest confidence. When you have completed all the questions, please return the questionnaire in the envelope provided.

AN INTRODUCTION TO THE QUESTIONS......

There are four main ways in which companies dispose of their waste. Please tick all the ways in which your company disposes of its waste:

- [ ] Collected by the local council (or their contractors)
- [ ] Collected by a private waste management company
- [ ] Taken to a disposal site by you or your staff
- [ ] Collected by salvage dealers or recyclers

From here on there are specific questions on each of these types of disposal. Please give an answer for **each type of disposal that you have ticked above**, including as much detail as you can. If you are not sure about the answer, please try to find out from colleagues who might know.

PART A: LOCAL COUNCIL COLLECTIONS

A1   Does the local council collect any waste from your premises?

☐ Yes   ☐ No **(If "No" please go to part B)**

For each type of container in the list below, please indicate how many containers are collected, how often they are collected, and typically what sort of waste goes into them.

A2   **Plastic Sacks/Dustbins (household size)**

How many do the council collect? .... Per ....

How full are they when collected? ....

What sort of waste goes in them? Please give details below:

	Type of Waste	% of Total in the Container
1		
2		
3		
4		
5		

A3   **Continental Containers (Eurobins)**

How many do the council collect? .... Per ....

How full are they when collected? ....

What sort of waste goes in them? Please give details below:

	Type of Waste	% of Total in the Container
1		
2		
3		
4		
5		

A4 **Paladin (cylindrical metal) containers**

How many do the council collect? .... Per ....

How full are they when collected? ....

What sort of waste goes in them? Please give details below:

Type of Waste	% of Total in the Container
1	
2	
3	
4	
5	

A5 **Skips**

How many do the council collect? .... Per ....

What size are they (give dimensions if not known)? ....

How full are they when collected? ....

What sort of waste goes in them? Please give details below:

Type of Waste	% of Total in the Container
1	
2	
3	
4	
5	

A6 **Other Containers (Please Specify)**

Type of container Size ....

How many do the council collect? .... Per ....

How full are they when collected? ....

What sort of waste goes in them? Please give details below:

Type of Waste	% of Total in the Container
1	
2	
3	
4	
5	

A7   **Any Other Loose Material (Boxes, Crates etc)**

Type of Loose Material	Size or dimensions	How much is collected?	How Often?
1			
2			
3			
4			

PART B: PRIVATE CONTRACTOR COLLECTIONS

B1 Within the last twelve months have you had any waste collected by a private contractor? *(Note: please exclude recyclers and salvage dealers - these are covered in Part D)*

☐ Yes   ☐ No   **(If 'No' please go to part C)**

B2 Please give details of your main contractor:

Name..

Address...

....

....

Carrier registration number....................................

Where is the waste taken to?...................................

....

For each type of container in the list below, please indicate how many containers are collected, how often they are collected, and typically what sort of waste goes into them.

B3 **Continental Containers (Eurobins)**

How many does the contractor collect? .... Per ....

What size are they? ....

How full are they when collected? ....

What sort of waste goes in them? Please give details below:

Type of Waste	% of Total in the Container
1	
2	
3	
4	
5	

B4 **Paladin (cylindrical metal) Containers**

How many does the contractor collect? .... Per ....

How full are they when collected? ....

What sort of waste goes in them? Please give details below:

Type of Waste	% of Total in the Container
1	
2	
3	
4	
5	

B5 **Skips**

How many does the contractor collect? .... Per ....

What size are they (give dimensions if not known)? ....

How full are they when collected? ....

What sort of waste goes in them? Please give details below:

Type of Waste	% of Total in the Container
1	
2	
3	
4	
5	

B6 **Other Containers (Please Specify)**

Type of container .... *Size* ....

How many does the contractor collect? .... Per ....

How full are they when collected? ....

What sort of waste goes in them? Please give details below:

Type of Waste	% of Total in the Container
1	
2	
3	
4	
5	

B6 **Other Containers (Continued)**

Type of container .... *Size* ....

How many does the contractor collect? .... Per ....

How full are they when collected? ....

What sort of waste goes in them? Please give details below:

Type of Waste	% of Total in the Container
1	
2	
3	
4	
5	

B7 **Any Other Loose Material (Boxes, Crates etc)**

Type of Loose Material	Size or dimensions	How much is collected?	How Often?
1			
2			
3			
4			

PART C: DISPOSING OF WASTE YOURSELF

C1 Within the last twelve months, have you taken any waste from your premises to a disposal point yourself? (Note: includes employees taking your firm's waste)

☐ Yes   ☐ No **(If 'No' please go to part D)**

Disposal site used..

Type of transport used ....

Type of waste taken ....

Weight or volume taken on an average trip ....

How often are trips made? ....

PART D: WASTE MATERIALS RECYCLED

D1 Out of all the waste your premises creates, are any items taken from your premises **to be recycled** rather than disposed of?

☐ Yes   ☐ No **(If 'No' please go to Part E)**

If yes, please give details below:

Type of Material Recycled	Amount (please give details of weight and/or volume)	How Often Taken or Collected?
1		
2		
3		
4		
5		
6		

Name of recycler or salvage dealer ....

....

D2 Have you included this recycled waste in the figures for Questionnaire Parts A, B or C **(Tick every box that is true)**?

☐ Yes, included in Part A

☐ Yes, included in Part B

☐ Yes, included in Part C

☐ No, this waste is additional to waste in A, B & C

PART E: GENERAL INFORMATION ABOUT WASTES

[*Note: if the WRA has any other questions, for example relating to duty of care, waste minimisation, it should include them here*]

FINALLY...

E1 Finally there are some brief questions about your firm or organisation.

(Please note that these details are needed for the analysis of responses and will be treated in strict confidence)

Name of firm or organisation ....

Address ....

....

.... Postcode ....

Name of person completing this form..

Position held ....Phone number..........................

Number of employees on these premises: ....

Full time ....   Part time ....

If you would like more information on managing your wastes please tick here •

Thank you for taking the time to fill in this questionnaire

APPENDIX E

Example postal questionnaire for larger firms

SURVEY OF WASTES

This survey is about the wastes generated by your premises – that's the premises to which this letter and questionnaire have been addressed. You may be a business or another type of employer (public, private or voluntary). You may produce a lot of waste or only small quantities. In either case, please fill in all sections that are appropriate to your organisation.

This survey has been designed to be quick and easy to complete. It consists of several preliminary questions and two types of forms to complete [mention additional booklet here if this has been included].

Most questions can be answered by ticking the appropriate box or writing in the space provided. If you do not know the answer, try to find out from a member of staff who would know.

The purpose of this survey is to enable us to estimate how much waste of different kinds is generated by different sizes of company in each industry sector. The results will be used to plan for the future to help ensure continuity of waste management outlets for waste from your business. It will also add to information on the national waste database – a database of information on waste arisings that you can access to compare your waste production figures with those of your industrial sector generally. This will help you see whether you might be able to save money easily by producing less waste.

The survey concentrates on weight data wherever these are available because they will provide you and us with a much more accurate measurement of the quantities and therefore the cost.

This is part of an official survey on behalf of [WRA]. If you have any queries, please contact [xxx] on [telephone number]. Thank you for your co-operation.

Remember: the cost of waste is not only what you pay for its removal, it is the cost of the materials that are lost.

Please be assured that all information provided will be treated in the strictest of confidence.

When you have completed all the questions, please return the questionnaire in the envelope provided.

A INFORMATION ABOUT YOUR ORGANISATION

(Please note that these details are needed to assist in the analysis of the information and will be treated in strict confidence)

Name of organisation ....

Address ....

.... Postcode ....

Name of person completing this form..

Position held .... Telephone number ....

Number of employees: Full time ....Part time ....

B WASTE COLLECTION/DISPOSAL METHODS USED BY YOUR ORGANISATION

Please put a tick next to each type of container or method of disposal your premises uses to dispose of its waste.

Sacks/bins ☐ Paladins ☐ Continentals ☐ Small open-top skips ☐ Large closed skips ☐ Rolonoff container ☐	Loose material ☐ Tanker collection ☐ **FILL IN FORM #2**
	Taken to disposal site by staff ☐
FILL IN FORM #1	**FILL IN FORM #3**

INSTRUCTIONS

For each type of container or disposal method you have ticked above, please fill in one copy of the appropriate form. For each tick you have in the column on the left, fill in one copy of the form that is marked FORM #1; for each tick you have in the right hand column, fill in one copy of the form that is marked either FORM #2 or FORM #3.

However, if you have the same type of container located at different places on your premises and containing different types of waste (eg paladins containing office waste at one location and paladins containing canteen waste at another location), please fill in one of the forms for each waste stream. If you run out of forms, please photocopy them or telephone [contact] on [telephone number] to ask for more.

SURVEY OF WASTES   PRODNO

Please write clearly in the spaces provided or tick the appropriate box

1. TYPE OF CONTAINER
   (See cover for examples)

2. GENERAL DESCRIPTION OF THE WASTE AND WASTE CLASSIFICATION CODE

3. DESCRIPTION OF THE PROCESS(ES) GENERATING THE WASTE

4. DETAILED DESCRIPTION OF WASTE

Waste Components	% of total in container
1	
2	
3	
4	
5	

5. PHYSICAL FORM OF THE WASTE

Please tick one only

- Solid
- Liquid
- Sludge
- Powder
- Gas

6. IS THIS WASTE ROUTINELY PRODUCED AT YOUR PREMISES?

Yes   No

7 IS THE WASTE SPECIAL WASTE?

    Yes   No

8 WHAT PROPORTION OF THE WASTE IS WASTE PACKAGING? %

9 NUMBER OF CONTAINERS COLLECTED/EMPTIED   Per

10 SIZE OF CONTAINER

11 HOW FULL ARE THEY? %

12 IS THE WASTE COMPACTED ON SITE?

    Yes   No

If yes, the compaction Ratio   :

13 NAME OF WASTE CARRIER

14 CARRIER REGISTRATION NUMBER

15 METHOD OF TRANSPORT

- Road
- Rail
- Inland waterways
- Other

16 DISPOSAL/RECOVERY METHOD

- Treatment
- Recycling
- Landfill
- Incineration

Other

17 WHICH SITE IS THE WASTE TAKEN TO?

18 OVER THE NEXT FIVE YEARS, WILL THE QUANTITY OF THIS WASTE

    Increase   +   %

    Stay the same

    Decrease   -   %

19 DO YOU KNOW THE WEIGHT OF THIS WASTE?

                     Yes   No

IF YES, WHAT DOES IT USUALLY WEIGH?

                   Per

LOOSE MATERIAL   FORM #2 PART 1

1. Description of material and waste classification

2. Process(es) generating the waste

3. Is this routinely produced by your premises?   Yes   No

4. What proportion is waste packaging?   %

5. How much is collected?   Per

6. Is the waste compacted on site?   Yes   No

   If yes, compaction ratio   :

7. Name of waste carrier

8. Registration number

9. Method of transport
   - Road
   - Rail
   - Inland water
   - Other

10. Disposal method
    - Treatment
    - Landfill
    - Incineration
    - Recycling
    - Other

11. Which site is it taken to?

12. Over the next five years will the quantity of this waste:

    - Increase   + %
    - Stay the same
    - Decrease   − %

13   Do you know the weight of the waste? ☐ Yes   ☐ No

What does it usually weigh? _____   _____

TANKER COLLECTIONS   FORM #2 PART 2

1   Type and size of tanker _____

2   Description of waste and waste classification _____

3   Process(es) generating the waste _____

4   Is this routinely produced at your premises?   ☐ Yes   ☐ No

5   Is the waste classed as special?   ☐ Yes   ☐ No

6   How much is collected?   _____ Per _____

7   Name of waste carrier _____

8   Carrier registration number _____

9   Disposal method
   - Treatment _____
   - Landfill _____
   - Incineration _____
   - Recycling _____
   - Other _____

10   Which site is it taken to?
   _____

11 Over the next five years will the quantity of this waste:

Increase ▢ + ▢ %
Stay the same ▢
Decrease ▢ − ▢ %

12 Do you know the weight of the waste?

▢ Yes ▢ No

What does it usually weight? ▢ Per ▢

MATERIAL TAKEN TO DISPOSAL SITE BY STAFF   FORM #3

1 Description of material and waste classification ▢

2 Process(es) generating the waste ▢

3 Is this routinely produced at your premises?
▢ Yes ▢ No

4 What proportion of the waste is waste packaging? ▢ %

5 How much waste is taken? ▢ Per ▢

6 Is the waste compacted on site? ▢ Yes ▢ No

If yes, compaction ratio ▢ : ▢

7   Type of site used

Treatment
Landfill
Incineration
Civic amenity
Recycling
Other

8   Name of site and location

9   Over the next five years will the quantity of this waste:

Increase   +   %
Stay the same
Decrease   -   %

10 Do you know the weight of the waste?

Yes   No

What does it usually weigh?

Per

STRICTLY CONFIDENTIAL

INFORMATION ABOUT THE COMPANY   DATE

Day   Month   Year

Company name

Address

Postcode

Contact name   Telephone No

INFORMATION ABOUT THE HEAD OFFICE (IF APPLICABLE)

Head Office name
Address
Postcode
Contact name   Telephone no

INFORMATION ABOUT THE BUSINESS ACTIVITY

Nature of business

INFORMATION ABOUT EMPLOYEES

Number of employees	Full time	
	Part time	
	Total (FTE)	

IN-HOUSE WASTE HANDLING FACILITIES

	Yes	No
Do you have any on site waste handling facilities or treatment plant?		

If no, go to the next page

If yes, which of the following facilities do you use?

Effluent Treatment		Material Reuse/Reclamation	
Boiler/Incinerator		Waste Reduction Unit	
Other (please specify)			

If No, but you intend to invest in or expand any facilities, please state which

Effluent Treatment		Material Reuse/Reclamation	
Boiler/Incinerator		Waste Reduction Unit	
Other (please specify)			

WASTE ARISINGS - (WRAs please note guidance will be required for waste producers to fill in this section)

1 Description of Waste	2 Process Generating the Waste	3 Type of container	4 Components (%)	5 Size	6 Number collected per	7 How full are they?	8 Compaction ratio
			1 2 3 4 5		Per	%	
			1 2 3 4 5		Per	%	
			1 2 3 4 5		Per	%	
			1 2 3 4 5		Per	%	

WASTE ARISINGS - (WRAs please note guidance will be required for waste producers to fill in this section)

9 Weight of the waste per annum (if known)	10 Waste Carrier	11 Carrier Registration Number	12 Method of Transport	13 Disposal/ Recovery Method	14 Name of Waste Management Facility	15 % increase or decrease in quantity over next five years

APPENDIX F

Data coding frame and form

Table F1 Coding Frame

PRODUCER NUMBER	Producers should receive a consecutive number, ensuring that there is no duplication of numbers within each area. This should be assigned prior to the sample being selected.
Group reference	Enter the letter of the group from which the company was selected (A-E) Missing = X
Interview method	1 Personal visit 2 Postal survey 3 Other
Waste stream number	Use a consecutive number for each waste stream at this location. The waste stream is represented by each survey form #2.
Waste description	Standard Department of the Environment code
Process (text)	The textual description of the process given by the producer should be entered here Missing = XXX
Process code	Since there is currently no standard process listing, regulators may wish to develop their own.
Component code	Standard Department of the Environment code
% (percentage)	Should be expressed as decimals (eg 90% should be entered as 0.9) Missing = 0.0
Mixed?	This will denote whether the components are mixed together. 1 Yes 0 No 9 Missing
Weight per annum	The weight of the waste per annum (if known) should be entered here, although only the number (eg 400 tonnes = 400) Missing = 0

Weight units	1	tonnes
	2	tons
	3	kilogrammes (kg)
	4	pounds (lbs)
	5	hundredweight (cwt)
	9	missing or not applicable
	Should any other unit of measurement occur, it should be converted to one of these	
Form code	1	Solid
	2	Liquid
	3	Sludge/slurry
	4	Powder
	5	Gas
	6	Other
	9	Missing
Routine?	This refers to whether the waste is routinely produced by the firm. This will become important in the data analysis.	
	1	Yes
	0	No
	9	Missing
Special?	Is the waste special waste? For postal surveys, regulators will need to apply their own expertise, or alternatively ring the company	
	1	Yes
	0	No
	9	Missing or not sure
Packaging?	Is the waste mainly waste packaging?	
	1	Yes
	0	No
	9	Missing
Container type	1	Sack/domestic bin
	2	Wheeled bin
	3	Paladin
	4	Continental
	5	Skip (open-top type)
	6	REL or FEL
	7	Rolonoff
	8	Tank
	9	Drums or barrels
	10	Loose material
	11	Other
	99	Missing
Number	Enter the number of these containers on the site	
	0	Missing or not applicable

No per annum	Multiply the number of containers by the number of collections occurring per annum eg 3 containers collected 2 times a week = 3 x 2 x 52 = 312 0 Missing or not applicable
Size	Only the number should be entered here as the units are to be entered in the next field, for example a 14yd^3 skip would receive an entry of 14. 0 Missing or not applicable
Size units	1 yd^3 2 m^3 3 ft^3 4 inches^3 5 cm^3 6 gallons 7 litres 9 missing Should any other measurement unit occur, the size of the container should be converted to one of those listed here.
How full?	This should be entered as a decimal (eg 50% = 0.5) 0.0 Missing
Compaction	The left hand number of the ratio should be entered here (provided that the right hand number is 1). If the waste is not compacted on site, then the ratio is 1:1 and a 1 should therefore be entered. A nought should never appear for this entry. (eg 3:1 = 3, 5:2 = 2, 1:1 = 1) Missing assume 1:1
Carrier (text)	The name of the carrier should be entered in words Missing = XXX
Carrier reg no	This is the carrier's registration number under the Registration of Carriers legislation. It will be in the form of three letters denoting the authority of registration with six numbers denoting the carrier, for example CAM030405 Exempt = EXEMPT Missing = XXX

Transport method	1	Road
	2	Rail
	3	Road and rail
	4	Water
	5	Pipeline
	6	Other
	9	Missing
Disposal method	1	Landfill - direct
	2	Landfill after pulverisation
	3	Landfill after milling
	4	Landfill after composting
	5	Incineration (on-site)
	6	Incineration (off site)
	7	Chemical treatment (on site)
	8	Chemical treatment (off site)
	9	Transfer station to landfill
	10	Transfer station to incineration
	11	Transfer station to treatment
	12	Recycling (on-site)
	13	Recycling (off-site)
	14	Borehole injection
	15	Mine shaft
	16	Lagooning
	17	Pyrolysis
	18	Composting
	19	Other
	99	Missing
Disposal site		Regulators should enter their own code for each disposal site in their area (eg licence number). For sites outside their area, the WRA of the site should be contacted and a number acquired from them. Where waste is being sent to other companies as part of a waste exchange programme, the company receiving the waste should be regarded as the disposal site and a number acquired from the WRA responsible for that company. Missing = 0
Site area code		Standard district codes should be used for disposal sites located within the regulators area. Where waste goes outside the WRA area for disposal, the DoE standard county codes should be used. Missing = 99X

Expected change	This should be entered as a percentage figure. So, for example, a 5% decrease in arisings would be entered as 95 whereas a 10% increase would be entered as 110. Missing assume no change = 100
Date	The date should be entered as DAY-MONTH-YEAR (eg 2/3/95). This information should never be missing. For postal questionnaires, the date that they were returned should be entered here.
SIC	The entire SIC numeric code should be entered here (for example 10.101). The oblique stroke separating the two final digits should NOT be entered. Missing = 00.000
Employees	The full time equivalent should be calculated by adding the full time employees to half of the part time employees. Missing = 0
Producer district	The DoE standard district codes. Missing = 99X

Table F2 Example of a completed coding sheet

PRODUCER NO	5101
Group reference	B
Interview method	1

Waste stream number	3
Description	S96

Process (text)	Administration
Process code	4

	Code	%	Mixed?
1	S60	0.5	1
2	S61	0.3	1
3	S70	0.4	1
4			
5			

	0		9
	1		1
	0		1

Container type	3		2
No per annum	104	0.956	2
How full?	0.95		1

	Waste Watchers		GHH202020
	1		1
	5065		61K
	110		

	5/2/95		10.101
	560		61K

INDEX

Accuracy of estimates 7.14
Accuracy of survey – national determination 7.17
Acts cited see individual entries
Administration of postal survey 15.1
Administrative boundaries – relevance 4.5
Aggregation – regionally and nationally 3.3
Agricultural waste – uses 12.74
Agricultural wastes 12.66
Allfirms database file 13.5
Analysis of data 7.34
Analysis of household wastes 12.10
Annual overlaps of surveys 7.23
Appointments for visits 14.4
Archiving data from surveys 15.24
Area background in plan 4.16
Area required for landspreading livestock wastes Table 12.1
Arisings see Waste arisings
Arranging visits by location 14.2
Availability of draft plan 3.36
Avoidance of double counting for some waste movements 10.16
Balancing supply and demand 10.21
Basel Convention 2.13
Benefits to industry 9.17
Best environmental option (life cycle assessment) 4.36
Bias avoidance in postal survey 15.22
Bring systems (e.g. bottle banks) 12.15
Business as usual option in plan 4.31
Business database 6.7
Business population – information required 6.12
Casual staff – limitations of use 5.37
Census of employment 6.7
Census of waste management facilities 11*
   as a data source 11.2
Characteristics of dependent variable 6.15
Checking data 7.32
Civic amenity waste 12.7, 12.8
Classification
   of industry 6.20
   of the population to be surveyed 6.18
   of waste 6.47, 6.54
Clinical waste 12.56
Co-operation
   national 5.32
   regional 5.32
   with waste collection authorities 2.34
   with waste disposal authorities 2.35
Co-ordination 3.5
   of recycling 3.9
   with development control planning 3.17
   with HMIP 3.16
   with NRA 3.15
   with waste disposal authorities 3.13
Coding data 7.30, 16.12, Appendix F
Collected household waste 12.8
Commercial waste – meaning 12.20, Box 12.2
Commercial wastes 12.20
Commercially confidential information 3.63, 5.25
Comparison between visits and postal surveys 13.26, Table 13.2
Complementary approaches – waste movements study and producer survey 10.36
Completing gaps in information from postal survey 15.20
Completion of visit form 14.12
Computer use 16.1
Confidentiality 3.63, 5.25
Confirmation of appointments with industry 14.7
Consent of WRAs receiving waste 3.46
Consignment notes as data source 5.8
Consistency in plan contents 4.7
Construction and demolition waste
      data sources 12.30
      recycling 12.32
Construction wastes 12.29

Consultation
   Environmental Protection Act 1990 requirements 3.30
   on the plan 3.20, Box 3.1
   phases 3.24
   public 3.23, Box 3.2
Contaminated soils 12.44
Contents of plan Box 4.1, 4.2
Control of Pollution Act 1974 2.2
Controlled Waste Regulations 1992 12.4, Appendix A
Conversion of volume to weight 16.20
Correlation with previous surveys 6.25
Costs and benefits 9
Costs – discounting for comparison of options 4.30
Cross checking waste arisings and movements data 5.18
Cut and fill
   meaning Footnote 129
   treatment of Footnote 129
Data
   analysis 7.34
   checking 7.32, 16.15
   coding 7.30, 16.12, Appendix F
   collection methods for industrial and commercial waste 13.24
   comparison between WRAs 9.14
   financial or calendar year 13.30
   items for census 11.5, Table 11.1
   possible errors 16.18
   to be collected in survey Table 14.1
   manipulation 7.27
   protection 14.6, 15.25
   sources for
      census of waste management facilities 11.2
      consignment notes 5.8
      construction and demolition waste 12.30
   validation 16.14, 16.16
Database for survey 10.19, 16.3
   design 16.5
   of waste 6.42, 12.11
Data coding frame Appendix F
Data management 16
Deciding the survey approach 13.36
Demolition wastes 12.29
Density factors 16.34
Dependent variable
   characteristics 6.15
   for industrial and commercial waste survey 6.14
   sources 6.17
Design
   of database 16.5
   of forms 7.36
   of survey forms 14.9
Development plans 2.16
Development plans
   structure plan 2.19
   waste local plan 2.20
Disaggregation of waste management planning data 3.3
Discounted costs 4.30
Distribution of waste production per employee 7.20
Division of industrial and commercial population 13.3
Double counting – avoidance 10.16
Draft plan – availability 3.36
EC Regulation on waste shipments 2.12
Effects of
   future legislation 8.20
   increased costs 8.18
   wheeled bin on collected household waste 12.8
Employee numbers to sub-divide industry and commerce 13.56
Ending the visit 14.28
Engineering spoil 12.37
Environment Act 1995 1.3, 2.15
Environment Agency Footnote 6, 1.8
Environmental Protection Act 1990 1.3, 2.3
Errors in survey 7.17, 7.24
Essential data items from postal survey 15.18
Estimates – accuracy of 7.14

---

*"11" means chapter 11 – and so on.

Estimates – improvement 6.30
Estimation of waste 6.27, 12
Evaluation of options for plan 4.25
Excavated materials 12.37
Exceptions to weight record requirements 10.29
Exclusions from industrial and commercial wastes survey 13.9
Exempt sites – record requirements 10.8
Existing conditions – description of 4.19
Exports of waste – government policy 2.10
Extraction of minerals 12.63
Factors for waste densities 16.34
Factors influencing amount of household wastes collected 12.14
Farm survey 12.72
Feedback of visit findings 7.38, 14.30
Feedback to industry 7.38
Financial or calendar year data 13.30
Flow charts 10.18
Flows of waste 10.13
Follow up of postal questionnaires 13.53
Food processing wastes 12.69
Forecasts 8.4
Forecasts
    industrial and commercial waste arisings 8.5
    recycling and recovery 8.16
    household waste arisings 8.4
Foreword of plan – content 4.9
Form design 7.36
Framework Directive on Waste 2.6, Box 2.2
Freepost for postal survey 15.4
Frequency of
    investigation 5.9
    survey 9.1
Future changes in waste streams 14.23
Future developments 8
Future legislation – effects 8.20
Future position see forecasts
Grouping waste producers 13.4, 13.14
Guidelines for postal surveys 15
Guidelines for visits 14
Hazardous waste 5.7
Health care wastes 12.52
Hospital waste see Clinical waste
Household waste
    arisings forecasts 8.4
    future arisings see Forecasts
    meaning 12.4, Box 12.1
    factors influencing collection 12.14
    other types 12.18
Household waste dealt with at home 12.15
Identification of business population to survey 6.5
Identification of waste production process 9.25
Importance of waste production process 6.57
Importance of weighing waste 9.23
Imports – government policy 2.10
Imports – requirement to plan for 2.10
Improvement of estimates 6.30
Improvement of records at waste management facilities 10.26
Improvement of waste management 9.19
Inbound and outbound wastes 8.12
Incineration residues 12.47
Incinerator ash – proportions 12.49
Incinerator – pollution control residues 12.49
Increased costs – effects 8.18
Industrial and commercial waste
    future arisings 8.5
    data collection methods 13.24
    population division 13.3
    separation of producers 13.6
    profile 13.2
    sample selection 13
    survey 13
Industrial waste – meaning 12.24, Box 12.3, Appendix A
Industrial wastes 12.24, Appendix B
Industry
    benefits 9.17
    feedback 7.38
    sectors excluded from wastes survey 13.9
Information
    from multiple sources 5.24
    other sources 5.28

    reporting year 5.15
    required on business population 6.12
Inputs and outputs of waste facilities 10.12
Inspection of site during visit 14.15
Integrated strategies for waste management 4.40
Intensity of survey 9.6
Introduction of plan – content 4.13
Investigation
    frequency 5.9
    legal basis 5.3
    meaning 5.2
    methods 5.11
    scope 5.16
Investigation, background and issues 5
Items of data required 11.5, Table 11.1
Key waste arisings 12
Landspreading 12.70
Landspreading – area required for livestock Table 12.1
Large waste streams – separation 12.27
Larger producers 6.38
Legal basis of investigation 5.3
Life cycle assessment 4.36
Litter 12.4
Lognormal distribution 7.20
MAFF (Ministry of Agriculture Fisheries and Food)
    information 12.67
Mailing the postal survey 15.9
Manipulation of data 7.27
Meaning of
    household waste 12.4, Box 12.1
    industrial waste 12.24, Box 12.3, Appendix A
    investigation 5.2
Medical and veterinary wastes see Health care wastes
Medical waste see Clinical waste
Methods of investigation 5.11
Mine and quarry wastes 12.61
Mineral extraction 12.63
Mineral types Box 12.4
Minimising waste 1.12, 9.19
Modifications to plan 3.61
Movements of waste 10
Movements v production of waste 10.30
Multiple sources of data 5.24
National co-operation between waste regulators 5.32
National household waste analysis programme 12.10
National waste classification 6.54
National waste database 6.42, 12.11
National waste exchange 6.46
Non-controlled wastes 5.29, 10.20, 12.60
Non-statutory consultation 3.25
Number of postal questionnaires required 13.50, 13.65
Number of questionnaires for postal survey 15.23
Number of visits required 13.39
Objectives of survey 7.8, Table 7.1
Objectives of waste planning Box 5.1
Origin of waste – site records 10.32
Other household wastes 12.18
Other sources of information 5.28
Other waste streams – separation 12.28
Packaging waste Footnote 89
Packaging waste Directive – requirements Footnote 89
Part 1 visit form – completion 14.13
Part 2 visit form – completion 14.15
Part 3 visit form – completion 14.28
Period of plan 3.52
Phased mailing of postal survey 15.14
Phases of consultation 3.24
Pitfalls in the survey 7
Plan, waste management see Waste management plan
Planned waste management operations 8.3
Planning Policy Guidance Note PPG 23 2.21, Box 2.3
Planning the survey 7
Planning the survey 7.4
Plans – development 2.16
Possible errors in data items 16.18
Post consultation review 3.41
Postal questionnaire
    examples Appendices D, E
    follow up 13.53
    numbers 13.50, 13.65

sample selection   13.69
time required   13.48
administration   15.1
archiving data   15.24
avoiding bias   15.22
completing gaps   15.20
essential data items   15.18
freepost   15.4
increasing the response   15.21
logging responses   15.16
mailing   15.9
print numbers   15.23
questionnaire content and design   15.6, Appendix D, E
questionnaire destruction   15.28
questionnaire retention   15.26
staggering the mailing   15.14
PPG 23 (Planning Policy Guidance Note no 23)   2.21, Box 2.3
Pre-consultation   3.25
Preparation for visit   14.11
Previous survey data – correlation with   6.25
Print numbers for postal survey   15.23
Probability distribution of waste production   7.20
Producer inspections – special waste   5.7
Producers – larger, survey method   6.38
Profile of industrial and commercial population   13.2
Proportion of waste to incinerator ash   12.49
Proportion of waste to incinerator pollution control residues   12.49
Protection of data   14.6, 15.25
Public consultation   3.23, Box 3.2
Publicity   3.33
Questionnaire
	content and design for postal survey 15.6, Appendix D, E
	destruction after postal survey 15.28
	retention periods for postal survey 15.26
Random sample selection for survey   13.59
Reclamation and recovery – identification and recording   14.16
Recording visits to special waste producers as inspections   14.29
Records of waste inputs   10.2
Records required for exempt sites   10.8
Recycling   2.34
Recycling
	and recovery – forecasts 8.16
	construction and demolition wastes 12.32
Reference numbers for waste producers on database   13.5
Regional co-operation for survey   5.32, 7.5
Regional co-ordination   3.7
Regional surveys   5.13
Reporting year   5.15
Resources – shortfall   13.44
Response to postal survey and increasing it   15.21
Responses to postal survey – logging   15.16
Results verification   16.39
Retention periods for postal survey questionnaire   15.26
Returns from waste management facilities   10.4
Review of requirements for survey   9.11
Reviewing draft plan post consultation   3.41
Role of Secretary of State   3.48
Role of waste management planning   2.27
Rural visit rate   13.34
Sample listing for survey   13.71
Sample selection
	for industrial and commercial waste survey 13
	for postal questionnaires 13.69
	random 13.59
	size 7.15, 7.18, 13.19
	systematic random 13.59
Sampling interval   13.60
Scope of investigation   5.16
Secretary of State's role   3.48
Selection of a waste management strategy   4.39
Selection of firms for survey   13
Self-sufficiency   2.13
Separate report on special waste   4.46
Separate reporting of types of waste   12.2
Separation of
	large waste streams 12.27
	industrial and commercial waste producers 13.6
	waste streams 12.27
	other waste streams 12.28

Sewage sludges   12.57
Shortfalls in resources – dealing with   13.44
SIC 1992 (Standard Industrial Classification)   6.21, 13.6, Appendix B
SIC – sort level   6.23
Site records of waste type   10.32
Site returns   10.4
Size bands of industry and commerce   13.56
Size of sample   7.15, 7.18
Small sample sizes – Student's t   7.19
Sorting the business population at SIC class or sub-class   6.23
Sources of dependent variable   6.17
Sources of information – MAFF   12.67
Special waste producer inspections   5.7
Special waste – volume   4.46
Spreading wastes on land   12.70
Spreadsheets   10.19, 16.3
Staffing   5.36
Staffing – casual   5.37
Staggering responses by phased mailing for postal survey   15.14
Standard Industrial Classification   See SIC
Start of the visit   14.13
Statistical methods   7.14
Strategies – integrated   4.40
Strategy – selection   4.39
Street sweepings   12.4
Structure plan   2.19
Student's t   7.19
Summary of plan – need and content   4.11
Supply and demand – balancing   10.21
Survey
	annual overlap 7.23
	classification 6.18
	data items Table 14.1
	deciding 13.36
	dependent variable 6.14
	errors 7.17, 7.24
	estimation 6.27
	of farms 12.72
	forms design 14.9
	frequency 9.1
	of industrial and commercial waste 13
	of industrial and commercial waste – size bands 13.56
	of industrial and commercial wastes – exclusions 13.9
	intensity 9.6
	objectives 7.8, Table 7.1
	planning 7.4
	population identification 6.3
	random selection 13.59
	regional co-operation 7.5
	requirements review 9.11
	sample listing 13.71
	sampling interval 13.60
	waste factors 6.27
	planning regionally 5.13
Survey of waste arisings   6
Time limit for statutory consultation   3.45
Time required for handling postal questionnaires   13.48
Timing of waste movements study   10.33
Types of mineral   Box 12.4
Types of waste to be reported separately   12.2
Tyres   12.3, FN118
Urban visit rate   13.34
Use of computers   16.1
Uses of agricultural waste   12.74
Validation of data   16.14, 16.16
Verifying results   16.39
Veterinary waste   see Clinical waste
Visit
	appointment 14.4
	confirmation 14.7
	ending 14.28
	feedback 14.30
	completion of form 14.12
	future changes 14.23
	grouping by location 14.2
	opening discussion 14.13
	part 1 form 14.13
	part 2 form 14.15
	part 3 form 14.28

Visit (*cont.*)
    preparation   14.11
    questionnaire   Appendix C
    rate for rural areas   13.34
    rate for urban areas   13.34
    reclamation and recovery   14.16
    recording inspections of special waste producers   14.29
    site inspection   14.15
    waste quantities   14.20
    waste streams overlooked   14.26
Visits and postal surveys – comparison   13.26, Table 13.2
Visits – number required   13.39
Volume to weight conversion   16.20
Waste arisings – defined   Footnote 35
Waste arisings – estimation   12
Waste classification   6.47
Waste collection authorities – co-operation with   2.34
Waste exchange – national   6.46
Waste exports – government policy   2.13
Waste facility – inputs and outputs   10.12
Waste factors – use in estimation   6.27
Waste flows   10.13
Waste Framework Directive   2.6, Box 2.2
Waste input records   10.2
Waste inputs v waste arisings   10.30
Waste local plan   2.20
Waste management facilities – census   11
Waste management facilities – weighing requirements   10.28
Waste management facility records – improvement   10.26
Waste management improvement   9.19
Waste management operations – planned   8.3
Waste management paper 25   12.55
Waste management plan   4
    background   4.16
    business as usual   4.31
    conclusions   4.42
    consultation   3.20, Box 3.1
    contents   4.2, Box 4.1
    evaluation of options   4.25
    existing conditions   4.19
    foreword   4.9
    introduction   4.13
    methods used to obtain data   4.17
    modifications to   3.61
    period covered   3.52
    publicity   3.33
    summary   4.11
    waste management strategy   4.33
Waste management planning – development   2
Waste management planning – role   2.27
Waste management planning – the process   3
Waste management strategy in the plan   4.33
Waste Management Information Bureau   12.11
Waste Management Licensing Regulations 1994   Footnote 2
Waste minimisation   1.12, 9.19
Waste movements   10
Waste movements
    outside area   see Inbound and outbound wastes
    study and producer survey – complementary approach   10.36
    study – timing   10.33
Waste movements study   10
Waste origin – site records   10.32
Waste packaging – information requirements   Footnote 89
Waste planning objectives   Box 5.1
Waste producers – grouping   13.4, 13.14
Waste producers – reference numbers   13.5
Waste production per employee – distribution of   7.20
Waste production process – identification   9.25
Waste production process – importance   6.57
Waste quantities – determination on visit   14.20
Waste Shipments Regulation   2.12
Waste streams often overlooked   14.26
Waste streams – separation   12.27
Waste type – site records   10.32
Waste types   12.3
Waste types – separate reporting   12.2
Wastes from food processing – categorising   12.69
Wastes – non-controlled   5.29, 10.20, 12.60
Weighing waste   5.31, 6.31, 13.21
    at waste management facilities   10.28
    exceptions   10.29
    importance   9.23
Wheeled bin – effects   12.8
WRAs – data comparison   9.14